Science and an African Logic

Science and
an African Logic

Helen Verran

The University of Chicago Press

Chicago and London

Helen Verran taught at Obafemi Awolowo University in Ile-Ife, Nigeria, between 1979 and 1986. She is currently senior lecturer in the Department of History and Philosophy of Science at the University of Melbourne.

Chapter 6 was originally published as Helen Watson, "Learning to Apply Numbers to Nature: A Comparison of English Speaking and Yoruba Speaking Children Learning to Quantify," *Educational Studies in Mathematics* 18 (1987): 339–57. © 1987 by D. Reidel Publishing Company; reprinted with kind permission of Kluwer Academic Publishers. Chapter 9 was originally published as Helen Watson, "Investigating the Social Foundations of Mathematics: Natural Number in Culturally Diverse Forms of Life," *Social Studies of Science* 20 (1990): 283–312.

The University of Chicago Press, Chicago 60637
The University of Chicago Press, Ltd., London
© 2001 by The University of Chicago
All rights reserved. Published 2001
Printed in the United States of America

10 09 08 07 06 05 04 03 02 01   1 2 3 4 5
ISBN: 0-226-85389-6 (cloth)
ISBN: 0-226-85391-8 (paper)

Library of Congress Cataloging-in-Publication Data

Verran, Helen.
    Science and an African logic / Helen Verran.
        p. cm.
    Includes bibliographical references and index.
    ISBN 0-226-85389-6 (cloth : alk. paper) — ISBN 0-226-85391-8 (pbk. : alk. paper)
    1. Yoruba (African people)—Science. 2. Yoruba (African people)—Mathematics. 3. Philosophy, Yoruba. 4. Ethnoscience—Nigeria. 5. Logic—Nigeria. 6. Ethnomathematics—Nigeria. I. Title.
    DT515.45.Y67 V47 2001
    160'.89'96333—dc21
                                                                2001027752

⊚ The paper used in this publication meets the minimum requirements of the American National Standard for Information Sciences—Permanence of Paper for Printed Materials, ANSI Z39.48-1992.

Contents

Part Four: Certainty

Acknowledgments

*M*y students in Ife are first to be acknowledged. The book is theirs as much as mine.

The research committee of Obafemi Awolowo University, Ile-Ife, provided financial support in two grants, which enabled this work. Colleagues in the Institute of Education provided encouragement and support. I specifically acknowledge contributions from friends Akin Aboderin, Joe Beddu-Ado, Barry Hallen, Segun Osoba, 'Diran Taiwo, Agbo' Folarin, Margaret Folarin, Karin Barber, Caroline Dennis, Morris Paul, Doreen Paul, Glenda Dare, the late Buki' Osibodu, and Pa Laosebikan. Joseph and Grace Odere ran my home along with their own and gave me time to think.

At Deakin University in Australia, I owe thanks to my colleagues in science studies and in the Faculty of Education, and to the Koori, Yolngu, and Yapa Aboriginal students who continued the work of my Yoruba students in alerting me to the issues. The History and Philosophy of Science Department at the University of Melbourne has provided support for the project over ten years. My graduate students contributed a great deal: Carol Steiner, Diane Mulcahy, Martin Gibbs, Cass Wrigley, Rey Tiquia, Jonathan Wearne, Margaret Ayre, Helen Smith, Chris Shepherd, Geraldine Cheney, and especially Katayoun Sadghi Rad Hassall.

The first draft of the book was completed in 1997 during a year's study leave in the Philosophy Department, Smith College. I acknowledge the support and resources that were placed at my disposal. To Kathryn Pyne Addelson, who arranged my stay, I am grateful, both for that and for her philosophy.

There is a group of people who have remained staunchly supportive through the long years of my struggle to think this through: Karin Barber,

David Turnbull, Anni Dugdale, Rosemary Robins, Bruno Latour, Donna Haraway, Lucy Suchman, John Law, Murrray Code, and Lorraine Code. I drew a great deal from their continued interest.

Finally my family. Husbands, Gabriel and Max; son, Daniel; daughter, Ruth; foster son, Jack; sister, Kaye; and granddaughter, Cecilia, have all had their lives made more difficult by the project of this book and have all made a unique contribution to it.

Part One: Introduction

Chapter One: Disconcertment

Could African numbers be different from the numbers of science? My experiences of struggling to teach science to Yoruba school teachers in Nigeria led me to ask that strange question. This first introductory chapter describes some of those experiences.

Chapter Two: Toward Generative Critique

Having elaborated the difference between Yoruba and scientific numbers, I was disconcerted to recognize that I had explained away the experiences that led to my odd question. I had seen real differences in numbers being managed but to tell of that I needed to find a new way to tell the realness of generalizing logic.

Chapter One

Disconcertment

*M*r. Ojo had a wonderful rapport with children. He relied on this in his teaching, often not bothering to gather much equipment for his science lessons. This morning he had excelled himself in his preparations, yet my heart sank when I saw how he had set himself up to teach the lesson on measuring length. He had assembled about twenty small cards: thick cardboard ten centimeters long and about five centimeters wide, marked off in one-centimeter divisions along the length; one card for each group of two or three children. To go with this, he had twenty lengths of string about two meters long. It was a lesson we had prepared as a group back in the laboratory in the Institute of Education at the University of Ife (now Obafemi Awolowo University). We had begun with one of the "Measuring Ourselves" pamphlets[1] produced by the African Primary Science Program. This program was part of a large and prestigious Western aid program that sponsored curriculum development in science and math with a focus on practical work in many countries in Africa in the 1960s and 1970s. Discussing how the lesson might be modified, so that it would be suitable for a group of fifty or so Yoruba children in classrooms almost devoid of resources, helped students prepare for their practical teaching exercise in the schools around Ile-Ife in southwestern Nigeria.

Like most of the students at the Institute of Education, Mr. Ojo was around my own age, and a far more experienced teacher than I was.[2] After completing primary schooling, these students had attended teacher-training colleges, a lesser form of Nigerian secondary school, from which they emerged at age seventeen or eighteen to take up positions as primary school teachers. They had held those positions for at least ten years before becoming eligible for two years of retraining at the Obafemi Awolowo University at Ile-Ife, after which they often taught in secondary schools. As a lecturer in the institute, I was responsible for at least part of their "retraining," but my inexperience when it came to Nigerian classrooms meant that this course in science education was, by necessity, very much a two-way program of training. As a group, students and lecturer, we worked out a way of negotiating the curriculum we were developing. The classrooms that these students taught in of necessity entered our negotiations. I taught solely in English; government policy and my meager capacity in Yoruba dictated that. The students too were officially obliged to teach in English in the classroom, but for the most part, and with my encouragement, they used at least some Yoruba in their lessons, engaging in so-called code switching as they felt the need. As part of our group preparation of classroom lessons, we spent time discussing how particular notions might be explained to the children in Yoruba. Necessarily I was silent in these discussions, following the students as best I could.

The lesson Mr. Ojo was to teach, "Length in Our Bodies," involved children using string to record another child's height, leg length, arm length, and so on, and then a meter ruler to report the length in metric units. In the lab, we had measured each other: use string to represent height, lay the string on the floor and use chalk to record the length, and then, when one of the few meter rulers becomes available, measure the distance between chalk marks and record the measurement in a chart.

We had also devised a means for evaluating the effectiveness of the lesson: Given a chart with fictitious children's names and heights, and using the process of the lesson in reverse, children would show a particular height, using a length of string to demonstrate whether they had understood. The students were nervous about teaching this way. It meant getting children out of their desks and putting materials other than pencils and exercise books into their hands. It meant children talking to and working with one another instead of working only from the blackboard and speaking only in reply to the teacher. The children were liable to become unruly and noisy at such a departure from the norm, and this could be a serious problem with forty-five to fifty in a small, enclosed space.

Mr. Ojo had, it seemed to me, taken an easy way out in lesson preparation with his little cards and a chart up on the blackboard.[3] I was not
impressed with his little cards, but I was soon forced to change my opinion. Speaking in Yoruba, Mr. Ojo demonstrated the procedure, completely leaving out the neat "lecture" on length that we had collectively
prepared back in the institute laboratory. (Under my instruction, we had
come up with a little account of length as a quality that things had: length
as infinite and so on.) Mr. Ojo called a small boy to the front. Amid the
giggles of the boy's classmates, he placed the end of the string just under
the boy's heel and, stretching it, held the string at the point that matched
the top of the boy's head. Tying a loose knot at this point, he took the
other end of the string from under the boy's foot. Holding this at one
end of a card, he wound the length of string around until he came to the
knot. As the boy retreated and slid down behind his desk, Mr. Ojo instructed: "Count the number of strings around the card. Ọ̀kan, èjì, ẹ̀ta,
ẹrin, àrùún, ẹ̀fà, èje, èjo, ẹ̀sán [one, two, three, four, five, six, seven, eight,
nine]. Write down the number. Multiply by ten. How do we multiply by
ten? Àádọ́ràán [ninety]." Then holding the bit of string remaining against
the graduations on the card: "mọ́kọnléláàádọ́ràán, méjìléláàádọ́ràán, mẹ̀tà-
léláàádọ́ràán, mẹ̀rìnléláàádọ́ràán [ninety-one, ninety-two, ninety-three,
ninety-four centimeters]. Yes, we have ninety-four centimeters. 'Diran's
height is ninety-four centimeters." I was scandalized. Mr. Ojo was presenting a bundle of short strands of string, a plurality, as length instead
of demonstrating the prescribed singular extension. The notion of extension, said to be essential for children to grasp as the "abstract" element
of length, seemed to have been rendered secondary, if not entirely deleted.

The children set to work in pairs or threes; soon the chart Mr. Ojo
had drawn on the blackboard was full of names and numbers. Several
children very efficiently used card and string to show how tall the fictitious
'Dupe, Tunde, and 'Bola of our evaluation exercise were. As a pedagogical
performance, the lesson could only be judged a complete success. The
children obviously had grasped using metric units to express a value. They
were pleased with their accomplishment. Mr. Ojo was proud, certain of
his success, and so was I. Yet at the same time I was profoundly confused
and puzzled. I had glimpsed the lesson as both the same as and different
from the one we had prepared, but I was unable to coherently explain
how it was either the same or different, or why discerning this mattered.
I was disconcerted and felt disempowered in being put out so. However,
this was not a power play against the òyinbó (white) lecturer on Mr. Ojo's
part. As he saw it, the lesson was a triumph for our group; I was included

as one of those who should feel proud at this little success; our lesson had worked well. He was disconcerted by my failure to evince whole-hearted and enthusiastic approval for his lesson.

Several others of my student group had placements in this school for their practical teaching round. When I went to watch their "length lessons" the next week, they too produced cards and string, adopting Mr. Ojo's technique with minor alterations. Soon all the teachers in my classes were teaching it this way—adopting the technology and using it with or without our little lecture on length. The lessons were routinely successful, and as a group the students swelled with pleasure at the success of their teaching in this new way of putting materials into children's hands. Generously, they ascribed their success to my interventions as they shared their newfound pleasure in teaching with me. As I saw things, the lesson we had devised from the African Primary Science Program cards had been quite transformed, with the demonstration of linear extension compromised in this way of teaching. For the students, it was still the same lesson: children learned to use metric units to measure and took pleasure in their learning.

The little cards evolved into many forms over the next few months, as they found their way into regular teaching practice. Ten-centimeter strips of tin; regular thirty-centimeter wooden rulers cut into three pieces, the original numbering erased, and numbers 1–10 reinscribed; bamboo pieces; and plywood strips with or without lugs to prevent the wound-up string slipping. For a few weeks, the technology became a craze among the children. String and card could be seen in many pockets. In school-yards, little knots of children could be seen stretching string, solemnly winding, triumphantly announcing the measurement, and eventually the winner. I came across children at a roadside stall measuring the yams they were selling; one enterprising boy had measured something very long, stringing four measuring cards together. However, the craze did not last long. The little cards disappeared both from the playground and the classroom, and failed to reappear during the teaching practice session the next year. Meter rulers again took up their accustomed place at the center of teaching length.

What was "out of order" in Mr. Ojo's lesson? Why did I experience this routine as disconcerting? I was aware then, and I still am, that when I try to tell others that something was out of order here, I am likely to be greeted with a shrug of the shoulders, a "So what? Surely it's just the same in the end, isn't it?" Perhaps it is. Yet it seemed this way of numbering had what I felt to be the essence of length almost as an afterthought.

Did it matter that some Yoruba teachers had children measuring length beginning in a *plurality* (a bundle of strands) rather than the *unity* of extension (a stretch of string)?

I was quite unable to explain the intense feelings that would well up in me when I watched my students teach measuring length with this little card and its many winds of string. Even now as I write about this episode, remembering my confused feelings of delight and suspicion, failure and success, I am shaking again with a sort of visceral laughter, the same sort of chuckling that often afflicted me as I watched my students teaching lessons like this one. I feel like I "saw" length and numbers, those serious abstract entities by which we organize so much of our modern lives, and "saw through" them at the same time. This double seeing never failed to get me. The numbering was familiar and other at the same time. The reality of number was momentarily realized as contrivance, but, oddly, this recognition of artifice did not seem to detract from its realness. Instead, it brought with it an "intense pleasure—the catching of the breath, the delighted laugh, the stirring of curiosity."[4]

I experience it also when I see the embroidered Kuna *mola* (richly decorated appliqué blouses made by Panamanian Kuna women) with its "copy" of the His Master's Voice copyright image set alongside the original in Taussig's *Mimesis and Alterity*.[5] "Pulling you this way and that, mimesis plays this trick of dancing between the very same and the very different. An impossible but necessary, indeed an everyday affair, mimesis registers both sameness and difference of being like, and of being Other."[6]

Staying true to this laughter eventually has me coming up with a new account of numbers, supplementing and revealing the familiar, conventional explanations of number. My contention, so passionately held that it motivates my long struggle to write this book, is that this laughter, which can easily turn to a visceral groan, this disconcertment, source of both clear delight and confused misery, must be privileged and nurtured, valued and expanded upon. These fleeting experiences, ephemeral and embodied, are a sure guide in struggling through colonizing pasts, and in generating possibilities for new futures. As a storyteller (a theorist) I treasure these moments, I do not want to explain them away. They are the first clue in my struggle to do useful critique. It is easy to ignore and pass by these moments—part of the problem is their fleeting subtlety— yet it is possible to become acutely sensitized to them. Interruptions, small and large are what we, as theorists, must learn to value and use.

Developing a useful account of what was happening there in Mr. Ojo's classroom that did not explain away that laughter took me a long time.

Conventional wisdom would pass by those puzzling small moments in the life of contemporary Nigerian (Yoruba) classrooms, seeing them as irrational glitches. Yet because they challenged my assumptions about numbers and hence many of my certainties, they shook me, and this effect was magnified because no one else seemed to feel anything amiss.

I am interested in the generalizing logics that have life among a group of people who call themselves Yoruba and live mostly in southwest Nigeria. These people, probably around 25 million of them, use Yoruba language, sometimes with numbers, in many arenas of their lives. Many of them also often use English, again sometimes with numbers, in those same arenas, as well as in different sorts of endeavors. These have been African ways of going-on for several hundred years by now. In this dual focus, I understand myself as doing "symmetrical anthropology."[7]

In attempting symmetry, it is important to acknowledge an asymmetry that lies at the center of my work and this book. The book is about two everyday African logics—Yoruba and English language use and number use—and how they relate to science. It is written in academic English that feeds off and into the logic of everyday English usage. In effect, one everyday logic is translated into the terms of the other, or to put it in a way that better reflects the power relations here, the logic of "the other" is translated into the terms of "the one." That asymmetry, the differential power and privilege associated with the two logics at issue, matters. In asking what numbers are, I must address the question of how using particular discursive strategies to elaborate knowledge of (an)other people risks merely recreating and reconstituting the interpretive frame out of which the discursive strategies grow.

In the school of my next story, the girls' uniforms were dresses made from green broadcloth with a bright yellow trim around the sleeves and the square-cut neck. The boys wore fawn shorts, their green shirts trimmed with yellow collars and sleeve cuffs. As usual, the nonuniformity of the children's uniforms delighted me, and during the morning session I found myself diverted by looking for the most originally styled version. Mrs. Babatunde was preparing her class for the experiment they would carry out after the morning break: "Measuring Lung Capacity." There was talk of lungs and volume, blowing and breathing, with much switching between Yoruba and English. Just before recess, Mrs. Babatunde had the children push most of their desks to the wall, and she spent her recess arranging six workstations in the cleared room.

Dividing the children into six groups, Mrs. Babatunde ordered each group to stand beside a pile of equipment. This was the most demanding of our lessons on "Measuring Ourselves": it was messy, with large volumes of water, and used complex equipment. When students taught this lesson, I brought six one-liter plastic graduated cylinders from the university laboratory; the students collected the rest of the equipment. The children were restless with excitement; perching on the pushed-back desks, they watched Mrs. Babatunde's demonstration and explanation. She had a metal trash can of the sort commonly found in government offices, half filled with water, and a large plastic container filled with water to the brim. She turned the container upside down in the bin, explaining why the water did not flow out. Then she inserted a rubber tube into the inverted plastic container, and taking a deep breath blew into the tube. Air bubbled up in the bottle, displacing water. "Look," she said, "this is the volume of air I breathed out of my lungs." Now to measure it.

Withdrawing the plastic tube and placing her hand over the top of the half-empty bottle, she lifted it out of water in the bin and placed it upright on the table. In the laboratory when we had practiced this lesson, at this point we had used a measuring cylinder to measure a liter of water. Pouring directly from the tall thin cylinder into the half-empty container, we measured how much water it took to fill the bottle to the top. Thus, we had measured the volume of air breathed out.

Mrs. Babatunde did not stick to this routine. She produced a small jug and had the children count how many jugfuls it took to fill up the bottle. Then she measured the volume of the jug using a graduated cylinder. With that figure and the number of jugfuls, she had the children calculate the volume of the air she had breathed out. She completed her demonstration of the task by writing her name and lung-volume measurement in the chart she had previously drawn on the blackboard, using the unit of milliliters.

Joyfully, the children crowded around their equipment, eager to try. Water splashed over the floor, forming huge puddles in the potholes in the concrete and soaking various children (I imagined what it would be like doing this in a classroom with a mud floor). The children huffed and blew, and using one of the little colored plastic milk jugs Mrs. Babatunde had provided, and the graduated cylinders I had brought, they calculated the volume of their lungs. Satisfyingly, they inserted numbers into the boxes of the chart ruled up on the blackboard. Mrs. Babatunde spent most of lunchtime restoring the room to some order, sweeping the flood out the door.

I was impressed and pleased with the lesson and so too was Mrs. Babatunde. She was a competent and confident teacher. "But," I put to her, "there was one small thing." I was mystified over why she had introduced the jug into the measurement when it would surely have been easier to pour water directly from the graduated cylinder to measure the volume of water displaced by the air of the breath. Mrs. Babatunde seemed confused and astonished at my question. Her first response was that it had never occurred to her to do it that way. I reminded her that we had done it that way when we practiced it in the laboratory. She did not remember; she began to get flustered, feeling that I was pointing to a serious mistake. I tried to reassure her that I thought the lesson was fine. After she had thought about my question for a while, she gained more confidence to question my understandings. She insisted that the children would not have learned to measure volume if she had done did it that way. When I gently probed her on this, she did not shift; she was quite certain. I had a feeling that Mrs. Babatunde was correct in this assessment, yet I could not see why it should be so. The feeling did not quite fit with my observation that on some occasions Yoruba children had learned to measure volume without the introduction of jugs into the measuring routine, although the lessons had not been as smooth as this one.

Inserting a number-of-jugs-full between volume to be measured and graduated cylinder: it seems to be a redundancy. Anyone still working with the informality of the quantifying we often see in cooking, for example, might mistakenly introduce jugs into the teaching of formal quantifying. Surely, many would say, I should have identified the redundancy and instructed Mrs. Babatunde on the "correct" process for formal quantifying. Yes, I might have done that, but I knew that she would go on using a number-of-jugs-full as a translating mechanism. Her commitment to the children as learners exceeded her commitment to the procedures of formal metric measuring. She knew that these Yoruba children would more easily learn to measure through conjuring up "a plurality." The children would get it when she presented volume as several jugfuls.

Mrs. Babatunde had an interest in the discipline sought by the official mathematics curriculum, which says that measuring volume should be achieved through the prior notion of a singular uniform extension. The curriculum insists on a very specific sequence of small bodily gestures as the proper way to measure volume. Those who are authorities in the matter of quantifying, insist that *that* is the way to proceed. Mrs. Babatunde might suffer if she fails to instill in the children the specific, required routines, developing in them the proper bodily habits of enumeration. If she failed my subject because she taught incorrectly, she might loose her job. At the same time, Yoruba children embodied an important set of interests for Mrs. Babatunde. She knew that they might not get it if she presented volume in the prescribed way, the little routines, gestures with hands and eyes, words and chart, might not congeal as number, and then she will have confused and rowdy children on her hands. She will have a very tiring day and go home with a headache.

Introducing the jugs seems to be such a small and insignificant alteration to the routine. It took me a long time to work out that it was a reversal of the constitution of plurality and unity similar to Mr. Ojo's reversal. *That* was what was disconcerting me here. When I did recognize it, like Jean Lave, who investigated the multiple arithmetics of Liberian tailors[8] and the different ways women shoppers and weight watchers quantify,[9] I could not see these episodes as exemplifying something like "protonumber": a limited or informal version of number. Here was a set of routines, a form of ordering that people lived their lives through, and it needed to be recognized as such.

Lave, listening to and watching women buying with numbers in American supermarkets, and men managing their weight with numbers in weight-loss clinics, saw that in these places number is done differently. The "officially sanctioned" way of doing numbers that is enshrined in mathematics curriculum documents, backing up the work of primary schools around the world, did not penetrate into these sites. Similarly, perceptive Brazilian researchers have shown that Brazilian children do number differently, and get correct solutions more often, when they are selling in the streets than when they sit in their classrooms.[10]

In discussing the performances of American women in supermarkets, Jean Lave points to the ways that standardized forms, like the forms of quantifying and doing arithmetic accredited in primary school curricula, can be understood as "attempts to dominate the definitions of the situations of the use [of numbers]."[11] The standard forms are politics. This politics has Mrs. Babatunde getting upset and defensive when I query her use of the jugs. For this reason, it matters which version of numbers she "does."

Yet there is an immediate politics of doing volume with numbers. This lies in the children getting it or not—a politics of going-on together in a particular time and place. If the way I "do" numbers passes by those that matter here and now, if they fail to recognize the routines I perform as number, then I will have an immediate problem to deal with. Acting on what might be called an implicit reading of the situation, rather than the explicit reading that choreographed her performance in the university teaching laboratory, Mrs. Babatunde brings six little jugs to school.

On the day, the banality of introducing jugs undermined my insistence that numbering through volume must be done *this* way. It had me saying to Mrs. Babatunde, quite duplicitously for someone in my position, that the lesson was fine. Yet it *was* fine, because, despite there being several reasons why it does matter which version of measuring volume Mrs. Babatunde has the children perform, there is also a strong sense in which it does *not* matter. Playing with the two versions of number that we can intuit as conjured up in the two routines is a bit like playing with a holographic toy where one picture is contained within another: tilting one way it's this picture, tilting another way it's another. I can flick from one doing of volume to the other—doing it in the teaching laboratory at the university with one long thin measuring cylinder, or doing it in the classroom with several small jugfuls—it is still doing numbering through volume. And perhaps this is even more disconcerting: to feel equally strongly both that it does matter and that it does not matter which way it is done.

Watching my students teach often produced vague and confused discomfort, as in the stories I have just told. Yet occasionally a lesson was so evidently a failure or a success, it offered itself for immediate explanation. Leaping to a neat closed story seemed to offer certainty: the comforting possibility of protection against blame for the failure, or the possibility of sharing in the praise warranted by outstanding success.

Mrs. Taiwo's classroom seemed even more crowded than others; the desks in their straight lines completely filled all but the front meter of so of the classroom. The children's skins gleamed after their morning wash; the boys' heads shaven or near to it, the girls' hair short and neatly combed, or braided. This lesson was from the same series as Mr. Ojo's, but Mrs. Taiwo was always meticulous in her preparation. On this morning, I was watching her third math lesson for the week. Of her previous lessons, the first had been very much a teacher talk lesson—in English, telling stories in which she had been trying to sensitize the children to the idea of length, helping them conjure up pictures of roads with string spread around the classroom. In the second lesson, Mrs. Taiwo had instructed children on the mechanics of using rulers. Now in this lesson they were to measure various sections of their body—height, head circumference, and so on—and record the lengths in a chart she had prepared for them.

Her talk was mostly in English as she gave what seemed to me to be a clear description of length.

Mrs. Taiwo demonstrated the procedure we had devised together. Holding the ruler on the top of a girl's head, she dropped a piece of string from the ruler until it just touched the floor. Then with the girl holding one end and Mrs. Taiwo holding the place on the string where it touched the ruler, they stretched the string on the floor and marked each end of the length they held between them with a piece of chalk. Using a meter ruler, she counted up the centimeters between the chalk marks, and writing the girl's name in the chart, she recorded the number of centimeters.

After Mrs. Taiwo had set the children to work, chaos ensued for twenty to thirty minutes as the large number of children enjoyed themselves in the limited space; some wrote numbers on the chart, others danced around with string, chalk, and ruler. In the chaos, it was hard to see who knew what they were doing and who was uncomprehendingly mimicking the actions Mrs. Taiwo had shown in her demonstration. After some calm had been restored and the desks pushed back into their (nearly) orderly rows, it became obvious that very few had generated any numbers. The logic of the exercise had escaped almost all of the children.

Mrs. Taiwo was disappointed and disconcerted that it had not "gone properly." Together we puzzled about what had happened. She was inclined to blame the "stupid children" or bad teaching from their regular teacher. I was not so sure. I knew these children; I had seen them last year confidently handling number in Yoruba talk, doing mental calculations in "marketing stories." Somehow, the string, the chalk marks on the floor, the number sequence, and the metric units had just not come together for these children as numbers.

I needed to assess Mrs. Taiwo's lesson, to give a mark out of ten, and I was puzzled and put out at her failure to connect with the children. At least there seemed to be something tangible to point to—a failure. More than that, there was an explanation ready at hand in the "cross-cultural psychology of learning" literature. Mrs. Taiwo's failure, which of course was also my failure, fitted with the by then well-established cross-cultural psychological fact of the "universality of delays in operatory development"[12] in children of African and other "traditional" societies. In particular for example, Barbara Lloyd had shown that both elite and poor Yoruba children scored poorly in "conserving continuous quantity"—like measuring with length.[13]

There was a huge and complex literature on this, complicated by competing theories of cognitive development, and competing claims about the beneficial effects of schooling. Despite these differences, there seemed to be quite widespread agreement that African children and adults showed poor perception and recognition of, and an inability to use, what are generally pointed to as attributes, qualities, or properties. There seemed no reason to exempt Yoruba people from this. Indeed, Yoruba commentators had themselves commented on this "deficiency" in Yoruba knowledge.[14] This apparent failure of "traditional peoples" to grasp the qualitative aspect of quantifying was considered by psychologists to be a very serious failing, for according to the orthodox accounts of quantifying,[15] the abstract and universal nature of quantifying lies within this qualitative element. Quantifying without this element was taken as of extremely limited utility, and classified as "primitive measurement."[16] Mrs. Taiwo's "disaster" can be explained away with a causal story of dismal failure by all involved. Blame is apportioned to teacher and teacher educator, and more generally to the "primitive Africans." This response seems to be vindicated by the fact that sometimes teaching Yoruba children to measure length with the singular extension embodied in meter rulers *does* work.

The school of my next story was located on one of the narrow roads radiating off from the Ooni's palace in Ile-Ife. The road was usually unaccountably clear and clean, and the school grounds too, much smaller than usual, always seemed neat, rake marks showing in the dust when we arrived in the morning. On Mondays, the children were well scrubbed; by Friday, their navy blue uniforms would be drab and dusty. Mr. Ojeniyi was dressed as usual in a clean, crisply ironed white shirt.

These were older children, who were quiet and responsive to Mr. Ojeniyi, sitting in orderly rows, their exercise books open before them. A lesson on division was scheduled, and I knew from painful experience that these lessons could be excruciating as teachers and children struggled with the mechanics of long division, tying themselves in knots with "the carrying over" and the "bringing down of the next column." I was expecting something good today. I always enjoyed Mr. Ojeniyi's lessons; he was one of those people "turned on" by the aesthetic of math. Numbers gave him joy and pleasure, which he communicated easily to the children.

Mr. Ojeniyi began in English, but after a few sentences shifted to Yoruba as he got warmed up in his explanation. I closed my eyes to try to pick up the Yoruba better and to follow his explanation. I was expecting an account of division as some sort of "serial process," something like the reverse of multiplication, understood as serial addition. I was astonished to hear Mr. Ojeniyi identify division as definitive of whole number: "You will not understand a number unless you understand the many ways it can be divided." He emphasized the point by repeating it in English, a common strategy in these classrooms. I lost his line of reasoning in my agitation, and the Yoruba became opaque. I tried to calm myself and just listen. Looking around, it was obvious from the looks on the children's faces both that they understood and that they were keenly interested in what Mr. Ojeniyi was saying. I was the only one in the classroom not paying full attention to the teacher.

I relaxed, catching again the gist of Mr. Ojeniyi's reasoning. First of all he presented a Yoruba number and showed it as a multiple of twenty plus or minus various factors of twenty, in translating it into a base-ten English-language number. Then he did the same process in reverse; using English number names and the base-ten system, he converted it into a Yoruba number, using division into sets of twenty as the first and defining process. The children followed his explanation and then wrote in their books the series he had elaborated for each translation on the blackboard. After two more such demonstrations on the board, one a Yoruba number translated to English base-ten number, the other an English number translated to Yoruba, Mr. Ojeniyi wrote ten Yoruba and ten English number names on the blackboard, all between one hundred and four hundred. He instructed the children to make similar translations for those twenty numbers, and to my amazement, most of the children completed the exercise with some facility.

Then came the fun. Each translation could be done in more than one way, yet clearly some ways were more elegant than others. Mr. Ojeniyi asked for volunteers, and the children loudly suggested alternatives. Amid much laughter and shouting, children jumping out of their seats to rush to the blackboard to demonstrate an alternative, a generally agreed-upon best translation was gradually reached for each of the twenty numbers. All thought of serious focus on the process of division vanished in the delight of the game, yet the game was all about division of whole numbers.

I had been told of and read about the base-twenty Yoruba number system (see chapter 3), and I saw that this was an inspiration in Mr. Ojeniyi's brilliant lesson. I was not the first to see the brilliance of Yoruba children in their number manipulations. Lloyd had noted that "Yoruba conservation [of number] results are as good or better than those of American children [tested in the same way]. Comparisons with other African groups were favorable though Yoruba performance did not match the near-errorless results of Tiv children."[17] Yet, according to commentators such as Hallpike, such brilliance would not necessarily imply "a full understanding of number . . . [for] a true grasp of the logical foundations of number extends beyond mere arithmetic and is integrally related to measurement [i.e., conception of qualitative extension]."[18]

The unexpected success of Mr. Ojeniyi's lesson can be explained away as yet another example of failure by "primitives." By adopting a particular regimen as necessary if the universal "natural number" is to be properly represented, a standard of true natural number is established, which both enables and justifies the attribution of "primitive" and "false" to other representations. However, an alternative explanation is at hand. Mr. Ojeniyi's lesson might be alternatively explained as a heroic resistance. It could be said that Mr. Ojeniyi had engaged understandings of a logic of Yoruba numbering that he shares with the children, but which is entirely opaque to the òyinbó (white) lecturer. A story could be told of Mr. Ojeniyi and the children joining forces to offer a heroic resistance to the European colonizing inherent in the accredited Nigerian mathematics syllabus.

Differing Accounts of the Generalizing Logic of Numbering

These stories about doing numbers in Yoruba classrooms seem to point to the existence of what in the social sciences is known as a "public problem"[19]—"the problem of mathematics and science education in African schools." There are two, apparently equally valid, ways of characterizing that "public problem," each associated with a particular political analysis of "the African situation." Rightist and leftist politics identify quite different causes for the conceptual difficulties that can arise in teaching and learning mathematics and science in African classrooms.

Adopting one or the other of these has a theorist distributing praise and blame in particular ways, and not only to the student teachers featured in these stories and to their lecturer. The contrasting analyses also distribute praise and blame to African communities, and to their European colonizers. In epistemological terms, this is the familiar contest between universalism and relativism. Such situations as those as those I have just told of were the grist for both "realist" and "relativist" intellectual mills in the "African thought" debate, which raged from the 1960s to the 1980s.[20]

Two rather differently targeted critiques can feasibly be developed. Necessarily they mobilize different accounts of the generalizing logic of numbering. I turn now to a brief consideration of each. First, I present what might be called a rightist critique engaging a universalist epistemology that relies on established conventional standards. It tells an orthodox story of institutional power relations and effectively explains away the disconcertment evoked by these mathematics lessons, as failure and inadequacy. The lecturer and in turn the teachers are accredited with institutional power to teach the canon—in this case the warranted way to measure length, volume, and so on. The teachers are under the jurisdiction of the lecturer who has the right and the responsibility to tell them they were wrong—award pass and fail. This critique valorizes the institutional power of the lecturer as the agent of "development," and sees only failure in these stories: failure on the part of "Africans"—children, teachers, and whole societies—who have "primitive" numbers. It is also failure on the part of the lecturer, who did not uphold the standards of modern university teaching and insist on the teaching of length in the "proper" way.

Embedded within this is a universalist view of number. It accounts a generalizing logic for that familiar entity "natural number," which, although "abstract," is real. This logic points to a structure "hidden" in the physical world, or implicit in the universal human experience of the world. This epistemological tradition emphasizes the given structure of the physical world (or the common experience of it) as the foundation for knowledge, which is understood as symbolizing or representing that structure or experience of it. Numbers as abstractions in some sense contain the structure of the physical world, and numbers are represented by numerals, which are symbols. In this tradition, there are multiple versions of the postulated mechanisms whereby the connection between the physical (or its sensing) and the symbolic might be achieved. I call this cluster of explanations universalism, and exploring its many versions is not my interest here.

I turn to the logical positivists, a group of philosophically minded mathematicians (or perhaps mathematically minded philosophers) whose project flourished in the middle years of last century for a detailed account of this position.[21] I take Carnap's *Philosophical Foundations of Physics* as exemplary. Here we see numbers as an expression of a singular universal logic; taken to be a found, universal form embedded in the structure of experience of the physical world. Carnap identifies a hierarchy of three kinds of concepts in the logic of numbering: classificatory, comparative, and quantitative concepts. By "classificatory concept," he means simply "a concept that places an object within a certain class," what we might ordinarily call a proper name. In science, they are the concepts of the taxonomies of zoology and botany (cool blooded or warm blooded, for example), but in ordinary life the earliest words a child learns—"dog," "cat," "house," "tree"—are of this kind also.

"Comparative concepts" play an intermediate role between classificatory and quantitative concepts. As Carnap presents it, comparative concepts are the qualities or properties held by objects. They are naturally occurring abstract objects found "in" things. The third level of the hierarchy is the "quantitative concept." A quantitative concept is the extent of a quality; it expresses an objective value of a particular body.[22]

Carnap's underlying assumption is that the world presented to the mind by human senses is as bounded, spatial extensions that endure over time. "Things" defined by being situated in space and time are perceived by knowers and available for classification. The assumption has the knowable world as an array of spatiotemporal entities with varying extents of various qualities. Qualities are the first abstract entities of quantifying, and numbers are a second-order abstract entity. Numbers arise in the analogy between the extension of qualities held by the perceived spatiotemporal entities and the numeral sequence expressing linear extension recursively. Numbers can thus represent a real value of a perceptual entity through a given quality.

On this account of what numbers are, Mrs. Taiwo is the praiseworthy one. She is correct in her presentation of length. Mr. Ojo is wrong to present an image to the children where length is portrayed as a bundle of string, whose many strands can be counted to come up with a value for length. Mrs. Taiwo, who presents an uninterrupted straight line, is portraying the proper and only correct way to "do" length. Mrs. Taiwo's line will be sectioned in a subsequent action to contrive units, and units accumulated to arrive at a plurality, but the linear extension is understood to constitute a true image of how qualities really exist in the world, and measuring *should* begin in *that*. This is the one and only correct way to teach children to quantify with length, and it is as important to get hold of the ideal as to get the right answer. Mr. Ojo has the children getting the right answer in the wrong way, in a "primitive" way. Similarly, Mrs. Babatunde, who rendered capacity as several (jugs), subverted the true meaning of capacity (volume) as a unified three-dimensional extension. As for Mr. Ojeniyi, with his suggestion that division is constitutive of number, he has failed to communicate the correct linear image of the number array. This way of distributing praise and blame grows from a particular account of numbers—one consistent with a universalist epistemology.

The alternative critique? A relativist account of the generalizing logic of numbering would lead to an analysis that insisted on Mrs. Taiwo's lesson as failing to engage a logic with which the children were familiar. It would point to Mr. Ojo, Mrs. Babatunde, and Mr. Ojeniyi as "heroes," mobilizing an indigenous mathematics. This leftist critique epistemologically engages relativism. An account of numbering that would underlie this critique might give a central role to language. Instead of positing a prelinguistic experience of physical world as universal, this theory of knowledge points to social practices like linguistic methods, as the origin of categories in knowledge. Such a position would maintain that Yoruba and English languages create alternative starting points for numbering, resulting in different quantifying logics. The explanation presents numbering in English and in Yoruba as working in separate symbolizing domains. It implies that children might learn either one or the other, or sometimes both if they are profoundly bilingual.

On this basis, both Mrs. Babatunde and Mr. Ojo are correct and due for praise. We understand that they do a good job with using Yoruba quantifying in a scientific setting, using metric units as prescribed by the curriculum. Mrs. Taiwo comes in for approbation on the basis that she failed to recognize and acknowledge the logical domain of Yoruba number, but in a relativist analysis, this is a pedagogical mistake, not a logical failing. In assessing Mr. Ojeniyi's lesson, we might argue that his arithmetic shows the juxtaposition of two systems of symbolizing number, suggesting how each set of symbols might be translated into the other. His lesson can be judged both logically and pedagogically sound.

The public problem of mathematics and science education for African children is rendered differently here. On this account, the problem is not the children, their teachers, and the confused university lecturer, but the curriculum. Prescribing a course of study suitable for European children, the curriculum fails to recognize the intellectual resources of these Yoruba children's community. The curriculum that might grow out of this characterization of the public problem would have education for Yoruba children proceeding in two symbolizing domains. Bilingual programs would be prescribed for inducting children into the two separate symbolizing domains, the one working with Yoruba logic, the other with English logic. Officials would begin to worry about the deleterious effects of mixing the two domains, and mathematics and science educators puzzle about what "Yoruba mathematics" and "Yoruba science" might be.

This leftist critique presents itself as an appropriate anticolonizing move. It promotes liberation of the Yoruba community, beginning in its schools, which have been laboring under the burden of a European curriculum for over a century. Yet a would-be reformer arguing that different communities embed differing forms of generalizing logic in numbering is at a distinct disadvantage in developing her program. Almost everyone believes that mathematics and science are universal forms of knowledge. This almost universal belief in numbers' universality begins with the notion of "natural number" as a found real object, as "culture-free" as a rock. To promote the idea of a program of bilingual education that emphasizes induction into Yoruba logic as much as a scientific logic seems preposterous.

Some Dilemmas around Teaching Mathematics and Science in African Primary Schools

As a university teacher, I felt faced with an acute dilemma. If I insisted that my students teach by the book, I would be training them as failing teachers. This seemed a betrayal both of my students and their pupils. Alternatively, if I encouraged my students in their innovations, and in teaching in Yoruba in ways that made sense to them, I could be perceived by some to be betraying Nigeria, giving into "tribalism," which is often seen as the moral position of cultural relativism. For a while, the sort of behavior I was encouraging in these Yoruba classrooms was known as "indiscipline." The military regime in power at that time, in conducting a brutal War Against Indiscipline, certainly did not exempt the staff and students of universities from discipline.[23] On another level, it seemed that to allow these "wrong" conceptions to pass would be betraying my commitment to scientific thought and my passionately held view that good, creative science could make a difference in developing better futures for Nigerian communities.

Despite the dangers, anxieties, and confusions, in the end it was the trust I had in my students as teachers that determined my actions as their lecturer. In supporting the creative efforts of my students in their classrooms in teaching science and mathematics in Yoruba, I determined to take on the seemingly ridiculous task of showing "natural number" as culturally situated. I respected my students as teachers. I needed to encourage and support what I saw as their fine pedagogy, even if this meant questioning a set of understandings that seemed unquestionable: the established and solid, universalist account of the generalizing logic of numbering. To credit my students' performances, it seemed that I needed to argue for numbers as multiple and various and against the idea that a number was a simple, singular and universal entity.

As I saw it, this argument could only be made by *demonstrating* many forms of numbering. It seemed that I needed comparative accounts of the generalizing logic of Yoruba numbering and in the modern sciences, as expressed in English in contemporary mathematics and science curricula. By privileging *practices* in numbering as I told of the generalizing logics, I saw myself avoiding the pitfalls of relativism, while arguing for possibilities of plurality.

In my focus on practices, however, what I failed to recognize were my own practices, specifically practices of translation. Failing to notice the extent and nature of the translations involved in my relativist telling of the generalizing logic of Yoruba numbering thwarted my intention of discovering where the difference lay and how it was managed, and hence revealing a politics embedded in numbering. As we will we see in the chapters that follow, although I developed a persuasive argument over differing symbolizing logics, the assumptions in a relativist framing meant that the insights I elaborated were almost irrelevant to the workings of the Yoruba classrooms in which the project was grounded. The claims I made spoke only to a very small, elite group of theorists with interests in arcane epistemological questions. Not only that: I recognized that, should my account of logics be taken seriously by curriculum planners, the resulting curriculum would be both pedagogically and morally unsound. Those disturbing insights were well in the future, however, when I set out on my project of studying the workings of Yoruba numbers, how Yoruba children learn to quantify, and the logics implicit in Yoruba and English language use. I called my project developing a theory of many relative logics.

The Structure of This Book

In the four years after I returned to Australia from Nigeria, its disconcertments fresh in my memory, I began developing a book manuscript I entitled "Numbers and Things," by preparing and publishing several papers. Engaging an orthodox, although controversial, relativism, I sought to develop a relativist theory showing the existence of many generalizing logics in numbering. When I got to the end of the manuscript and tried, in conclusion, to elaborate what followed from my analysis, I began to see that the difference I wanted to focus on and keep was explained away.

My students had, it seemed, known how to manage differences in generalizing logics by contingently connecting and separating, but the difference I had described was not open to management. In that account of difference, you must do either one or the other; it is not possible to manage things by doing both at the same time. In a serious way, my analysis had failed. I was stymied. I abandoned the manuscript and began again. Eventually I came up with a method that had me overtly decomposing my previous relativist writings. That method determines the structure of this book.

In this chapter, I have told stories pivoting around the disconcertment I experienced in my participation in mathematics and science lessons in Yoruba classrooms. Chapter 2 continues the introduction, elaborating the disconcertment I experienced at finding that I had explained away exactly the phenomenon I wanted to describe and explain. The subsequent chapters are arranged in three parts: "Numbering," "Generalizing," and "Certainty." These parts deal with the workings of numbers, the generalizing implicit in using numbers, and the certainty that seems to characterize judgments made through numbering. Each part contains three chapters, and the sequence of chapters in each part is similar. The first chapter is a relativist account of difference in some aspect of the generalizing logics of numbering, the second chapter is a "decomposing" of that argument, and the third chapter develops an alternative account of difference in that aspect of generalizing logic.

A feature of this book is the way I decompose my own texts put together some ten years ago, as critique of universalist understandings. My decomposing should not be taken as repudiation on my part. I do not seek to deny those critical relativist texts with their originality/banality, any more than I repudiate the brilliant/ordinary lessons of my students. The sequence of chapters in each section does not constitute a redemption narrative. Instead, I struggle to keep the tensions. The relativist papers failed to deliver a useable critique; nevertheless, they are significant and important, and not only because they embody a considerable creativity and were born from dedicated scholarship. A generative critique offers the possibility of innovation, a way that things might be done differently to effect futures different from pasts, and innovation needs something to work on. Generative critique does not offer salvation, but it does offer possibilities of enlightenment. Accounts of numbering as founded symbolic logic are not going to fade away, but they can be used in more and less enlightened ways.

Chapter Two

Toward Generative Critique

"Seró, ókọn, éjì, ẹ̀ta, ẹ́rin, árùún . . ." (zero, one, two, three, four, five . . .). The chant, partly absorbed by the thick mud-brick walls of the classroom, streamed through the large gaps that served as windows. It mixed with the street vendors' cries and the bleating of goats as they passed along the road onto which the classroom windows opened. I stood on the side of the road, leaning into the classroom through one of the windows. A very crowded classroom this, apparently the outcome of the recent declaration of universal primary education in Nigeria. The tiny children clad in royal blue uniforms perched in twos and threes on wooden forms, their feet dangling above the packed-down earth floor. Each child had an exercise book open, but they were not working with pencils; instead, this was a lesson with spoken number names and bottle caps. Oddly, the children each had a largish pile of bottle caps before them. Mr. Adeleke explained to me later that his wife owned a beer parlor, so maintaining his collection was easy. He carried the caps around in a plastic bucket, carefully collecting them after each lesson.

The children moved a bottle cap from the left-hand to the right-hand page of their exercise book and then chanted a number name, evidently naming the value of the gradually enlarging collection of caps on the right. Mr. Adeleke instructed the children to sit up straight. He ensured that they carried out the hand movements in rhythm to the chant, and with a military precision. He stopped at *okòó* (twenty), urging the children to place their hands around the pile now sitting on the right-hand page. As they did this, he had them repeat the name, "Okòó." Then to the tune of the numbers counted backward, they moved the caps in the opposite direction. He interrupted the flow. Pointing to the right-hand pile, he asked how many were in the pile; children shot their hands into the air. A girl was chosen; she correctly named the number remaining on the right-hand side. Continuing until the last cap was removed from the right-hand page of the exercise book, pointing to the now empty page, Mr. Adeleke asked again how many were in that pile. This time only one hand went up. A boy answered, "Seró" (zero), triumph in his voice.

Satisfied, Mr. Adeleke issued instructions for the next chant; this time the counting was forward in twos. He began again like a sergeant major, "Seró." He stood to attention, pointing to an empty page. Looking around the classroom, he admonished a squirming child to sit still and point at the empty page when he said "seró." The barked "seró" seemed at once to herald the activity of counting and to name the value of the (nonexistent) pile. He commenced the chant again, "Seró, éjì, ęrin . . ."; each time the children repeated the name as they picked up two caps, transferring them from left to right.

Here was another dedicated and creative teacher of primary school mathematics. His use of materials to support the learning of number names and their sequencing was unusual; even more unusual was his inclusion of zero, apparently transliterated from English, as the starting point for counting a collection. This was not the modernized version of Yoruba counting sanctioned by the authorities.[1] It seemed a spontaneous synthesizing, at once emphasizing both a performative element often associated with Yoruba counting and a referential moment as prescribed in the modern primary school curriculum.

In other classrooms, the chanting of number names was used more like a crowd-control technique, more a version of the collective performance of putting hands on heads than an experience of numbering. Invariably that type of counting begins with *ókọn* (one). Mr. Adeleke's innovations indicated that he had given some thought to the matter of teaching Yoruba children the conceptual basis of numbering. At the same time, the lesson worried me. Was this syncretism heresy? Was it a form of "cultural pollution"? I remembered the warnings of educational psychologists on the dangers of simplistic mixing registers that might cause cognitive dissonance in bilingual learners.

Despite these anxieties, I felt that this lesson had a place alongside those of Mr. Ojo, Mrs. Babatunde, and Mr. Ojeniyi, as a brilliant yet puzzling success. It too showed up Mrs. Taiwo's disconcerting failure. Yet again, I found it difficult to name what was happening, and to evaluate its significance. In some ways, it seemed highly significant that Mr. Adeleke introduced a transliterated zero and had the saying of Yoruba number names linked to actions with hands, eyes, and material entities. In other ways, the innovations seemed entirely banal and of no importance. Is it all the same in the end? Do alternative ways of teaching, whether in English or Yoruba, with or without bottle caps, with or without zero, add up to teaching the generalizing logic of numbering?

As this lesson progressed, I took several photographs of Mr. Adeleke and his class of small pupils, each with their pile of bottle caps. I was at that time writing a series of articles about the exigencies of Nigerian classroom life for an Australian newspaper. A photograph of Mr. Adeleke, wearing a smart, pale brown safari suit and bending over a child apparently busy calculating, accompanied one of these articles. It is one of my many photographs of these students in action. I was proud of their accomplishments, as they were themselves, and these photographs recorded our small triumphs. With these photographs, I carried classrooms, children, and teachers back to Australia after I left Nigeria.

I remembered these photographs when I was struggling to complete the manuscript "Numbers and Things." Hoping to recover some of the feelings associated with being in those classrooms as inspiration, I displayed the photographs on the walls of my study in Australia. The effect of these photographs covering my walls was much greater than I had anticipated. Their display set off serious doubts about the significance of my neat schemas of the logics of numbering in English and Yoruba as they are laid out in this book in chapters 3, 6, and 9.

Sitting among those photographs, I saw that the satisfying contrasts of my papers fell far short of what I was aiming for. It seemed that my theory of logics as relative and plural had *explained away* the very phenomenon I wanted to point to. I intuited that in those classrooms, as in the wider sphere of contemporary Yoruba community life, things went along in ordered ways that were at once *both and neither* English logic of numbering and/or Yoruba logic as I had so assiduously laid them out. I had failed to grasp the generative tensions of my students' innovative and successful lessons. I suspected that the logical schemas I had described were like designs for machines that could never actually work!

In a flood of nostalgia, the photographs brought back Mr. Adeleke and his incorporation of *seró*, transliterated "zero," into the routine of teaching little children to count in Yoruba. Uncomfortably, I remembered my recently formulated argument on why the continuing absence of zero is what should be expected and perhaps even actively promoted with Yoruba numbering (see chapter 3). A mathematics education policy that might grow out of my elaboration of dual logics—the extension of my papers that I was just then setting out to develop—would outlaw such lessons. Promoting the learning of Yoruba logic in Yoruba schools for the "good of the Yoruba," it would have schools working to ensure the survival of a logically consistent Yoruba numbering system by both keeping it "pure" and ensuring that Yoruba youngsters were skilled in its use. Yet, looking again at that photograph of Mr. Adeleke, I remembered my strong sense of the rightness of that lesson. At the time, I was convinced that Mr. Adeleke was a talented teacher who knew what he was doing, and I remained so.

My project had begun in momentary appreciation of the possibilities my students demonstrated of different ways of doing things in science and mathematics lessons. Yet, in the papers that I had written and that I was now trying to turn into a book, this recognition that something odd was happening had become an argument for distinct knowing worlds achieved through the workings of alternative abstract logics. Distinct logics? Dual, abstracting conceptual systems? These photographs reminded me that my students did not think of their pupils' minds, nor of their own, as the resting place of these two logics, nor had I when I participated in the life of these classrooms. I had described a difference, but what had actually disconcerted me—the connections between different ways of doing numbers—could not even be imagined. The difference I had described could only be imagined as an inevitable separation; the capacity to do Yoruba and English number at the same time was inconceivable.

The effect of the photographs was to remind me of the puzzling inclusiveness of many of the lessons and the failures when this inclusiveness was absent. I went back to the few exercise books recording these lessons that I had brought back to Australia as mementos. In the notes on which I had based my assessments of the students' teaching, there were no signs indicating that my students were acting out of an appreciation of two different spheres of logic. On the contrary, like the photographs, the notes reminded me of the seamlessness of the lessons in which Yoruba teachers taught Yoruba and other Nigerian children the basics of modern mathematics and science, often speaking most of the time in Yoruba.

The photographs and the notes eventually formed the basis for the stories with which I began this book. I used those stories in chapter 1 to point to the interruptions that first started me puzzling, the first outcome of which was a series of relativist papers (chapters 3, 6, and 9). Actually, the writing of those stories grew from the disconcertment I am elaborating here, the realization that my relativist papers explained *away* the phenomenon I was trying to grasp. Writing those stories was a form of therapy by which I regained what I felt to be important in those lessons, and that writing began the second stage of my long struggle to tell the workings of generalizing logics in contemporary Yoruba life.

My Iconoclastic Critique

My project began with a desire to refute number's claims to social, cultural, and political neutrality. I had plunged into my empirical studies, angry at the way universalist accounts of number refused recognition of mathematics and science lessons as a form of political culture. Their universalist claims rendered as plain wrong all but one set of banal routines with hands, eyes, words, and other things like bottle caps. In that account of numbering, there is no choice and no politics, only truth set against falsity originating in technical incompetence. Chapters 3, 6, and 9 claim recognition for mathematics and science lessons as arenas of politics by showing alternative generalizing logics of numbering. That was an outcome of my relativist studies. The papers make some significant and persuasive arguments against universalism, establishing instead possibilities for many forms of the generalizing logic of numbering. They refute "established proof" of "primitive" African quantification and oppose the notion of deficiency in the African psychological makeup often supposed to be the cause of this "primitive" quantifying.[2]

With iconoclastic intent,[3] my relativist study set out to demolish the universal category of "natural number." As I saw the situation then, that universalist way of telling the realness of number needed to be destroyed because the object it justified—natural number—was a fetish, an illegitimate object that embodied a desire for certainty. "Natural number" as a singular, universal object was, as I saw it, animated by a psychological projection. It was an expression of a rationality needful of a certain and singular foundation, an object "number" on which to base an ordered world. In my eyes, this universalist "natural number" was a false idol, at once the basis and the product of a politics of domination and oppression—an imperial expression of the Enlightenment.

It seemed that truth telling, engaging the true spirit of the Enlightenment, would shatter the notion that number was a "given" of a universal experience and that all its instantiations exhibited a single logic. Truth telling demanded that the universalist account be replaced with a notion of number as founded on material orders variably achieved in practices like tallying goods. A plurality of logics understood as relative to each other would truly tell the realness of number. This telling could in turn be the basis for policy to institute a social order in which Western colonizing through school curricula might be held in check.

If I had been called on to elaborate here, I would have argued that the false idol—the universalists' natural number—constituted a basis for an "othering" explanation. The explanation legislated the primitiveness of Africans and established the need for their uplifting through development—modern education. Difference is ruled out in universalism as it legislates a particular and, I would have argued, an abhorrent moral order: "You should give up your nonmodern Yoruba ways to become full knowing subjects in the process of making modernity in Yorubaland!" For a universalist, Western colonizing is an agent of progress in Africa, and any notion of postcolonialism is neocolonialism—a continuing struggle to roll back the darkness through learning to use the given universal categories singularly embedded in material reality.

The alternative image of numbering that I argued for has social practices as the origin of a foundation for knowledge. Numbers originating in different forms of symbolizing is the norm. My project was linked to Bloor's program in sociology of knowledge,[4] which goes back to Mills's identification of originary ordering practices as the (social) foundation of knowledge. The picture of Yoruba- and English-language generalizing logics that the three relativist chapters of the book present (chapters 3, 6, and 9) has three sets of social practices as foundational: practices of generating number names in tallying, practices of dividing the material world into units, and practices of predicating and designating in ordinary language. Although I emphasized the practical nature of these routines, in the end I point to them as historically set social methods of constituting symbols.

My relativist explanation presents doing number in English and in Yoruba as separate logics of generalizing, carried out with abstract objects in symbolizing domains. This implies that children might learn either one or the other, or sometimes both logics if they are profoundly bilingual. It has foundations given in disparate, historically located, social practices, resulting in many worlds, where knowledges compete and where a deep incommensurability underlies a superficial capacity to translate between symbolic systems. This story contains the possibility that schoolteachers, as agents of local resistance, can oppose the university lecturer, an agent of Western imposition. It also legislates a moral order. It prescribes the heroism of resistance set against the tragedy of alienation. Colonialism here is oppression and destruction of the indigenous forms of knowledge by powerful Western forms of knowledge. An incommensurability separates "Western" and "other" knowledges. Postcolonialism in this account is the expulsion of invading Western forms by a renaissance and resurgence of indigenous forms of knowledge.

This oppositional stance implies a separation of Yoruba and English logics, and the need for vigilance in order to keep things that way. It suggests that children should be trained in both logics. The existence of these two logics opposing each other in the imposition of/resistance to colonizing suggests that a proper anticolonialist cultural policy here would be to maintain the purities. Yet what sort of education is that? An abhorrent apartheid of numbering has no place in mathematics education. What I had spent so much time and effort in elaborating—two logics—now seemed to be prescribing policies that were both wrong and appalling. It becomes apparent that the critique is morally irresponsible.[5]

Missing the Generative Tension

The papers I present as chapters 3, 6, and 9 were developed only after I had relinquished my position at Obafemi Awolowo University in Nigeria and returned to Australia. This separated me from my Yoruba colleagues, and consequently opportunities for presenting my work to this important audience have been very limited. On the two occasions when my audiences boasted a significant presence of native Yoruba speakers, I observed, much to my confusion, that the most significant response of my Yoruba colleagues was eyes twinkling in amusement, as if a collective laugh was always about to erupt. The second occasion on which I presented these formulations was in February 1994. A seminar, "Yoruba Number Counts!" in the Center for West African Studies at Northwestern University, Evanston, Illinois, was attended by several native Yoruba speakers. The story I told then used some of the puzzles generated by the children's responses (chapter 6), which I attempted to solve by grammatical analyses of the sort I make in chapter 9. They heard my talk out, the amusement evident on their faces. At the end of my presentation, my friend Agbo' Folarin, from Obafemi Awolowo University at Ile-Ife, in Nigeria, congratulated me. "It is amazing," he said, "that you found a way to say all that about Yoruba number. It just makes me want to laugh!" and he did! It disconcerted me to find my Yoruba colleagues erupting with laughter as, expecting praise and congratulations, I held up my "discovery." Why does my serious empirical investigation of Yoruba logic make native Yoruba speakers want to laugh?

Considering this unexpected response, I began to wonder how I might use the results of my relativist study were I to find myself back in the science education laboratory at Obafemi Awolowo University, and back in Yoruba classrooms. I felt strongly that its use would be very limited, pulled out only to justify to those in authority why I encouraged my students to teach mathematics and science in Yoruba. The notion of alternative numbers with differing logics is certainly not something I would have tried to explain to my students, or convince them about. I imagine the stares of incomprehension, the suppressed laughter should I have tried. They would have been embarrassed on my behalf. My ever-practical and grounded students would just politely ignore me, and rightly so, if I attempted to get them to teach, say, "English arithmetic" in one lesson and "Yoruba arithmetic" in another. It would clearly be foolish. "Number is just not that sort of a thing," the more outspoken might have explained.

What seems to follow from my relativist critique is that schools should teach the logics as separate. Imagine the unlikely circumstance that a curriculum authority took this seriously and set up a committee to develop a curriculum on this basis. Such a committee might begin in an uncontroversial way by prescribing bilingual education. They might move on to insisting on separate time slots for Yoruba mathematics and English mathematics: Yoruba arithmetic on Tuesday morning, English arithmetic on Thursday afternoon. They might then ponder the topics to be included in "Yoruba science" and in "English science." The scenario is both absurd and repellent. Such a curriculum is likely to both harm and handicap Yoruba learners. It would also generate boundaries within Yoruba life where, currently, generative connections predominate.

My relativist analysis carries the absurd implication that bilingual Yoruba knowers of logic and mathematics must in every circumstance choose consciously and deliberately to be one thing or the other, with "us" or with "them." It makes it appear that they must always decide which "bit of mind" to engage in the form of the questions "Which language/logic should I speak here?" "Which world should I be in here?" The inevitable dualism implied in the relativist critique separates Yoruba logic and English logic, rendering everything in contemporary Yoruba life a complete dichotomy.

Bilinguals do need to make choices about what are often called "registers," yet it seems that their decisions are not different in kind between similar sorts of choices made by monolinguals. Just as monolinguals often wittingly and unwittingly mix registers, so do bilinguals, combining Yoruba and English words, phrases, and sentences. The problem with my relativist "difference" is that the separation it implies is an ultimate division. Refusing this picture of contemporary Yoruba experience, I do not want to imply that opportunities for choice and deliberation over logic do not exist. That would deny mathematics and science as forms of political culture. I want to refuse the notion that the choice is internal to minds and expressed as individual decision over which symbolic schema to engage. I insist that there is a politics here, but I have difficulty in naming and describing that politics. What are the choices and where are they located?

As I felt my way toward what it was that my theory of logics as plural and relative had missed, I saw that foundationist accounts of logic, universalist and relativist alike, fail to recognize communities of practice as creative and generative. Yoruba teachers, and their pupils, and even their lecturers, make up an arena generating new ways to go on, and regenerating old ways of going on together. Foundationist analysis fails to recognize the creativity of collective life as a logical going-on in actual times and places. My schema misconstrues the generative nature of doing mathematics, and doing science in teaching and learning, as much as in other contexts. This misrepresentation hides the nature of the struggle that teachers and learners everywhere experience, but particularly so in situations like those of Yoruba classrooms.

It seems that my students made choices between banal routines; selections between and among what were little more than alternative, familiar sequences of gestures with hands, words, and stuff. Sometimes they deliberately joined the banal routines in innovative ways, and often the lessons that emerged when they did this were especially successful. They evoked numbers that children recognized as numbers while discovering within them new possibilities of orderings that might be effected by numbering. In saying this, however, I do not wish to imply that numbers are *really* processes. In those classrooms, numbers enjoyed a general recognition as entities, things of some sort. Mrs. Taiwo, in her failed lesson on length, for example, accepted that, for the most part, numbers were absent from the classroom on that day. Yet as entities, numbers are in no way pure, abstract, or symbolic, and they are not fragile. I might say that in those classrooms we all "believed" in numbers. Yet to say "I believe in numbers" seems a quite foolish statement, for numbers do not seem to be something that it is necessary to *believe* in. Numbers are familiars that seem to "do" us (make us who and what we are) as we "do" them.

There was a generative tension in those brilliant but unconventional math and science lessons of Mr. Adeleke and Mr. Ojo. It was a difference that enabled connection as easily as it allowed separation, and presumably it allowed both in multiple ways. This creativity is not confined to classrooms but is a characteristic of Yoruba community life in general, and not just Yoruba life. The generative tension over numbering (unusual because we do not often see such creativity around numbering) that I saw expressed in those Yoruba classrooms is a particular example of the creativity of collective life in all times and places. This familiar and banal experience is entirely missed by my relativist critique. I began to suspect that all foundationist analyses—universalist and relativist alike—must miss this phenomenon. It seems to be somehow outside the frame of that analysis. Yet to put things this way was indeed shocking, for I had not seen my analysis as framed. Naive as this sounds to me now, I had not thought of my relativist analysis as being framed apart from its framing by the world itself. I had thought of myself as telling things the way they were.

An Abhorrent Moral Economy Is Embedded in My Theory of Logics as Plural and Relative

Recognizing that I had missed the very phenomenon I sought to grasp in my elaborate description of two relative logics, I began to see also that a moral economy was implied in what I had developed. My critique carries with it a set of answers to the moral question "How should we live?"[6] I understood this gradually through first identifying a paradoxical element in my analysis. I had elaborated difference by denying it as real in the sense that it is a site of choice. Locating two different logics in dual conceptual systems with shared origins in generic human practices, I actually removed possibilities for doing difference by contingently separating or connecting.

At the same time, the boundary between "the Yoruba" and "the modern" had been hardened and solidified in my analysis as the categories were (re-)defined and delimited. A denial of difference as real or doable *and* a simultaneous hardening of separation are some of the moral outcomes of my relativist way of telling "a Yoruba logic." Recognizing this, I saw that as a foundationist analysis, my relativist schema merely retells an imperial universalism. Inverting its claims, it remakes the naturalness of colonizing power relations through opposition.

More shocking than this recognition, however, was seeing that the form of the two splendid objects I had "discovered" exactly fitted the form of objects that I had shown as participating in English language logic. The form of the abstract objects, "logics," was analogous to a spatio-temporal object—a thing. As we shall see later (chapter 9), this is the form of objectness found in English language usage, but not in Yoruba language usage. In my relativist analysis, Yoruba logic is revealed as a abstract object "in" Yoruba life which, remarkably, has the characteristics *not* of a Yoruba language type of object, but of an English language type of object, albeit in a rather peculiar form.

Another way to say this is to recognize that I had "discovered" Yoruba logic as a special case within English logic. The abstract object "Yoruba logic" I had revealed could only be taken as an echo, a shadowy form of English logic. The schema reenacts the categories of a universal modernity, originating in European traditions, and a Yoruba echo of a necessarily European modernity. Either way, a distinct "us" and "them" are locked forever together, and apart, through the specter of originality/mimicry. This is just what we would expect of a difference "outside worlds"—not real or doable. The only way to tell such a difference is to pull "their" world into "ours."

Perhaps most painful for me as a teacher and theorist was a third insight about the moral economy embedded in my relativist work. I recognized that, despite my being a significant, although puzzled, disconcerted, and often bumbling participant in the situations out of which my analyses grew, in my relativist writing that persona had quite disappeared. The author-in-the-text of chapters 3, 6, and 9 is a removed observer, an invisible presence outside and perhaps above the endless confusions of Nigerian classrooms. Further, my analysis grew from what I had experienced as rather confused messing around on my part. Now, despite being an absence in the texts, I find that, as author, I have assumed the voice of authority. This voice, legislating from a position of certainty, tells the ways contemporary Yoruba should understand themselves and their knowing.

Uniformitarianism

In the name of respecting difference and diversity, my theory promotes a covert form of "uniformitarianism"[7] by making difference unrealizable. This is the term used by A. O. Lovejoy in characterizing the Enlightenment belief in the unity of mankind and the uniformity of the laws of justice, morality, and aesthetics governing it. As he elaborates it, attributing it to post-Newtonian writers, it is a recognizable ideology. I am using the term to name a much more muted and underlying impulse in both forms of foundationist analysis. Even in these times when differentialism or diversificationism is the dominant ideology espoused in the academy and many parts of popular culture, a form of uniformitarianism still thrives through the workings of foundationism.

I argue that an unacknowledged, even denied, uniformitarianism is embedded in relativist analysis, and that the foundationist framing that universalism and relativism share is its origin. Foundationism of any sort is committed to ideals that are necessarily uniform. The denial of difference as real with its concomitant redefining and limiting, the rendering of "Yoruba logic" as a degraded from of that logic originating in Europe, and the reinscribing of the author as authority are all expressions of the unacknowledged idealism that infects relativism as much as universalism. Being unacknowledged, the impulse to legislate uniformity in relativist argument is more difficult to deal with. Foundationism is a metaphysics that denies it is a metaphysics.

Universalism, with its image of knowledge as symbolic representation referring to underlying givens of either the world or experience of it, brings with it a specific ontology that, with its framing imaginary, enables a particular realness. There objects are either abstract or material, but irrespective of that, they are *real*—with realness here implying *givenness.* A stone, for example, is structured separations: bit of real matter located in real space and enduring across real time—a material object. So is a number real, except number is structure either in matter or in the perception of it, in the givenness of human biology. Number is abstract. Relativism, with its image of knowledge as a system of categories emerging out of various schemas of symbolizing that build on social practices of working the material, similarly has objects as real in the sense of given (although this is contested by universalists). Here, however, stones and numbers are both beyond crude matter; both have been "worked-up," constructed in social life, although numbers more so. There is a sense in which all objects are abstract in relativism.

Foundationisms prescribe ultimate meanings—metaphysics—but fail to recognize this. I discovered the hard way that critique enabling a telling of difference in generalizing logics, which might be grasped by those whose lives and worlds are made with that difference, cannot be framed without challenging the self-evidence of the stories of realness embedded in foundationism's ontology and metaphysics.

A Contention

Where and how did I lose track of what it was that so disconcerted me? Why did my project of exposing universalist accounts of number seem laughable in the context of ordinary Yoruba life? I had sought insightful analysis that might inform practice both of teachers of mathematics and science and of a wider community of scholars, but I had produced a theory from which risible and abhorrent prescriptions followed. I had remade the boundaries I meant to dissolve. Seeking ways to respect difference, I had instead promoted a pernicious uniformity. How had this dysfunctional and unpalatable moral economy become embedded in my theory of logics as plural and relative? These questions crowded in on me. They need answers that are canny, robust, and generative if I am to have any chance of producing an insightful telling of the generalizing logic of any form of numbering in a way that avoids merely rehearsing the separations and divisions of established moral orders, remaking past as future. The sentiment animating my intellectual struggle to critique conventional accounts of number and logic is recognizably an Enlightenment attitude, but these painful recognitions forced me to accept that, to have a chance to pulling it off, I needed to escape an Enlightenment frame of telling.

My contention is that for generative critique we need a new story of realness, of how and where realness originates. This project will not deny its assumptions; it will acknowledge itself as a metaphysics. I can put this another way and say that we need to entirely reimagine the project of ontology.

I contend that the flaws I now recognize as marring my theory of logics as relative and plural originate in a set of unnoticed assumptions in my study. I worked with self-evident notions of the world, knower (mind), and knowledge. What these entities are and how they relate were entirely taken for granted. To put it bluntly, I literalized or fetishized the figure that has world, knower, and knowledge as given separate entities. I took up Enlightenment assumptions that are commonplace in the modern academy and in many everyday modern lives.[8]

Transgression of the boundaries of the given categories of world, knower, and knowledge is perhaps the most serious category mistake in foundationism. It is sometimes claimed that "primitives" transgress the boundary between knowledge and the world in literalizing—constructing fetishes by having symbols and objects as the same thing. This "mistake" subverts all claims that "fetishizing primitives" might make for the certainty of their explanations of the world.[9]

Let me spell out the metaphysical assumptions of my relativist analysis; it sounds so ordinary we wonder how it could be something as grand as "metaphysics."

> Worlds are physical, knowable orders of matter set against empty space-time. (Universalists and relativists disagree on the origins of that order, the first locating it in the physical, the second in past human work.)
>
> Knowledge is representation of abstract or ideal categories.
>
> Knowing is located in minds of removed, judging observers of order(s) in the physical world, who formulate knowledge.

I adopted the less orthodox version of this set of assumptions—relativism—the version that has the order that symbolizing knowledge represents as the outcome of past knowers' interactions with the physical world. Yet in common with my universalist colleagues, I took it for granted that the world, knowers, and knowledge are a priori separate and that these three given entities would feature in and constitute my explanations. The assumptions amount to what I have been calling foundationism, and are commitments that universalism and relativism share. They enable and allow the juxtapositions my theory made, but the cost of this is high. They add up to a curriculum I find morally and pedagogically unsound, and prescribe an answer to the question "How should we live?" that repels me.

To be more precise in elaborating my contention: It is *not* the configuring as such that led to the troubles I have outlined above. What made my relativist telling of the generalizing logics of Yoruba and English just another imperial prescription was my failure to recognize and respect this tripartite configuration of world, knowers, and knowledge as a trope, as a working figure. I took this modern frame for telling as literal in a simplistic and unreflective way. It never occurred to me to ask about this ordering, this familiar, comfortable, and solid configuration.

It is important to recognize that claiming this usefully configured modern constitution[10] as tropic or figurative does not imply that I doubt its reality. It amounts to a refusal to literalize the modern figuring of world, knowers, and knowledge, yet insists on its realness, effected in myriad doings of modern life. In belatedly noticing and problematizing this ordering with its tripartite separations, I certainly claim it as real. This realness is precisely why any critique of science that is going to be useful in our times and places must be produced without the aid of those comfortable, achieved separations. Please understand that I am *not* saying here that world, knower, and knowledge are sometimes separate and sometimes not. Or to say that another way, I am not proposing that at some times or in some places world, knower, and knowledge are given as separate, and that at other times or in other places the separation between world, knower, and knowledge is just a figure, a way of talking of something that really does not have those separations. That way of understanding my claim still retains the ordering of the figure of time or space in a literal way.

To describe this perplexing move is difficult. A significant problem is that saying it in the ways I just have in my previous paragraph does not help anyone master the tricks, as you must already understand. Yet it is possible to show it, and in this book I show it three times. In a transitional decomposing moment, my three relativist texts are worked toward quite different sorts of stories of the generalizing logic of numbering. To engage this move in critique, we must become familiar and comfortable with figure and reality as simultaneously both contrivance and context, or as Latour has it, as simultaneously fetish and fact—a "factish."[11] Imagining this realness is a difficult and tricky move for a modern thinker to master. The making of that move is a main focus in this book. In part it is about changing the metaphors and narratives that lie at the heart of stories about realness.

The greatest challenge in this is being patient enough to become familiar with the odd objects/subjects we then find ourselves dealing with. In the third chapters in each part, I tell stories of these odd new sorts of objects/subjects, each focusing on a differing aspect of the oddness. The new story of numbers highlights the multiplicity that appears to be "inside" objects like numbers in this frame. Recognizing this can be quite shocking, for we are used to numbers being singular, pure, and essential. In telling the realness of generalizations differently, I focus on the workings of what I call ordered/ordering microworlds as the very specific locations where the complexity of ongoing life is rendered as complicated objects and subjects in effecting generalizations. In the third new sort of story, I give a new account and location for the certainty we feel in numbering and generalizing.

Critique for Postcolonial Times and Places?

In the episodes related in chapter 1, I stand as a participant caught up in complex power relations. Being both teacher and learner and at the same time a theorist, a storyteller of the episodes, reciprocal indebtedness and accountability characterize my position. How might I fully acknowledge the complexity of this position, yet still do critique that might be useful?

To adopt either of the foundationist accounts of numbering is to foreclose and legislate. The explanations of the "problem" of science and mathematics curricula in places like Ile-Ife that grow from these accounts, and the solutions that follow from those explanations, remake a colonizing modernity. Distributing praise and blame, they reenact the categories of either a universal modernity (originating in European traditions) or a Yoruba echo of (European) modernity. Foundationist explanations *fail* the critical project. They actually make it impossible to imagine futures different from pasts. In developing critique that allows another "mode of relation to disconcertion,"[12] I want to avoid endlessly rehearsing old framings, yet allow the possibility of arguing/negotiating toward futures different from pasts.

Rather than explain away disconcertments, I want critique that keeps the puzzlement of sameness and difference we can see in Mr. Ojo's lesson, in which the little mysteries of Mrs. Babatunde's jugs and Mrs. Taiwo's rulers stay around. Keeping the delight and curiosity potent is important. I want to privilege the disconcertment. This involves imagining a way of doing critique that effects neither the dominating unity of universalism nor the oppositional fragmentation of relativism. Needing to attend closely to events/episodes and their re-presentation in story to tease them apart to reveal what is "inside," this critique acknowledges itself as transitional and transactional. It is transitional in the sense that it seeks to resite the issues, and transactional in the sense that it insists on claim and counterclaim both staying around. This sort of critique does not imagine itself as a second-order reframing that replaces.

The new form of critique I attempt in this book is distinguished by the framing implicit in my new stories of numbers, generalizing, and certainty. This implicit set of working images/stories tells realness as emergent: what's real emerges in gradually clotting and eventually routinized collective acting, and not only human acting. I call these framing images and stories "an imaginary," although I hesitated in settling on this term, for it can easily be misunderstood. Like many terms used in academic literature, imaginary has life in vernacular English as well as arcane meanings in various specialized discourses. My hesitation over using "imaginary" stems from the recognition that my usage seems not to overlap at all with popular usage. For me, the imaginary does not involve the mind and is certainly not located there. There is a significant disjunction between the imaginary as I mobilize it in this book, and "the imagination," which links with fantasy and is taken to be an aspect of the mind. My usage seems close to that of the cultural theorist Michael Carter: "The most fundamental distinction that can be drawn between the imaginary and the couplet, fantasy and imagination is the fact that the former is not a distinct form of mentation at all. It is not a 'thing' of the mind but an overarching relation. . . . The imaginary is not something which the subject calls up at will, or . . . slips into when the reality principle is lifted."[13]

For Carter, the imaginary is constituent of the very situation of any doing or action. He quotes Castoriadis in justifying his usage of the term. "[D]oing posits and provides for itself something other than what simply is, because in it dwell significations that are neither the reflection of what is perceived, nor the mere extension and sublimation of animal tendencies, nor the strictly rational development of what is given."[14] The imaginary implicit in my new telling of numbers, generalizing, and certainty links up with Castoriadis's account of any doing, and eschews the picture of a given foundation and its symbolizing, instead having worlds, with their objects and subjects, as accomplished in collective going-on.

Schemas picturing emergent categories feature in many explanatory arenas in contemporary intellectual life, for example, Charles Darwin's evolutionary schema in biology. So we need to be careful lest confusions creep in right at the beginning. The emergent categories that many evolutionary theorists focus on are the categories they take to be emerging in a given natural foundation: emergent foundational entities that knowing can be about. Theirs is a thoroughly foundationist (universalist) way of doing knowledge. So too is the imaginary of emergence implicit in some relativist metaphors of constructivism.[15]

In my new working images/stories, worlds emerge all of a piece. The imaginary refuses a priori separation of the symbolic and the material, although it recognizes that such category separation might be achieved.

This is to picture knowing as routine and necessarily embodied collective acting emergent in the life of particular times and places, contingently linked both within those times and places and with other times and places. Adopting such a framing has me dealing with and telling stories of real worlds, where objects and subjects materialize, or "clot," in particular figured ways. As objects/subjects, they emerge and participate in real worlds as relational. More or less obdurate associations are mobilized in their participation in collective going-on. Having worlds as emergent does not dismiss the fact that some worlds are done as foundationist worlds. On the contrary, it recognizes that as a considerable accomplishment and offers the possibility of generative critique to those worlds.

Colonialism in this frame of collective and relational emergence is the outcome of, and embodied in, ongoing collective acting. This does not need to imply that the colonizer and the colonized have the same understandings or shared points of view, that somehow colonizing is a cozy agreement between oppressors and oppressed. It goes beyond understanding responses to colonialism as either treacherous collusion or implacable resistance. Postcolonialism here is not a break with colonialism, not a revolution, a history begun when a particular "us," who are not "them," suddenly coalesces as opposition to colonizer. Nor is it a taking over of the colonizing agenda by the former colonized. In this narrative frame, colonialism is remade in postcolonial enacting. Postcolonialism is the ambiguous struggling through and with colonial pasts in making different futures.[16] All times and places nurture postcolonial moments. They emerge not only in those places invaded by European (and non-European) traders, soldiers, and administrators. Postcolonial moments grow too in those places from whence the invading hordes set off and to where the sometimes dangerous fruits of colonial enterprise return to roost.

In this book, I show a form of critique that I suggest is generative of postcolonial times and places. It is a form of critique I describe as transitional and transactional, and importantly, it contains a decomposing moment. As I see it, the three chapters of each part constitute a transitional refiguring in moving from a foundationist imaginary for telling realness to an imaginary of emergence. The middle chapter of each section decomposes by revealing how the foundationist critique depends on literalizing. The decomposing move in each section is transitional in that difference is "resited" or relocated. No longer in symbolizing, difference in the second sort of story is located in banal routines of ongoing acting. In this transition, the decomposing chapters are transactional. The transitional chapters hold those that precede and succeed them in tension, undoing each other by mutual revelation of their figuring.

Anthropological Inspirations for Critique through Framing Worlds as Emergent

Marilyn Strathern came to appreciate the possibilities offered by imaginaries of emergence through taking seriously the accounts and explanations offered by her Melanesian friends. In her writings,[17] she moves back and forth between the imaginaries of foundation she finds, say, in British parliamentarians debating reproductive technologies, and the imaginaries of emergence, which suffuse the ways Melanesians tell reproduction.[18] She notes that the sorts of ethnographic approaches that take worlds as emergent, refusing the category distinction between the material and the symbolic as a priori, are the sorts of studies for which anthropology is renowned. She insists that they are important for contemporary research in fields far beyond anthropology.[19]

I take the term "decomposing" from Marilyn Strathern's elegant article "The Decomposition of an Event."[20] She identifies "interrogative decomposing" as a form of analysis indigenous to Papua New Guinea Highlanders: "[W]hat we might call analysis in Hagen takes the form of decomposition, taking apart an image to see/make visible what insides it contains. . . . Forms appear out of other forms, that is they are contained by them: the container is everted, to reveal what is inside. We can call this indigenous analysis. It follows that past and future become present: any one form anticipates its transformation, and is itself retrospectively the transformation of a prior form."[21]

As Strathern tells it, in the Highlands of Papua New Guinea such decomposing analysis is also a passing over to those who participate as analysts. In ceremonies where Hagen men put on their grandest displays of wealth, while it is they who dance and sing, it is the audience who must do the work. The audience decomposes the presentation and recomposes, in this work recognizing their responsibilities as analysts. Such a "showing what is inside" can never be exhaustive, nor does it lead inexorably on.

I am not pretending that the notion of decomposing is in any way unique or new in academic discourse. It is entirely conventional in disciplines like literary and cultural studies and is orthodoxy in postcolonial studies, where it is more commonly called deconstruction and applied to textual compositions, rather than events/episodes in collective life.

In the Yoruba literary forms of *oriki* (praise poetry) and *ìtàn* (formalized storytelling), and the ways they work together, I suggest that we see the sense of decomposition that Marilyn Strathern points to in Hagen analysis. I use Karin Barber's *I Could Speak until Tomorrow: Oriki, Women, and the Past in a Yoruba Town,* as my reference here.[22] While Barber does not explicitly identify oriki as a critical form, in reviewing her book Toyin Falola suggests that the following point to oriki as a critical literary form: "The hidden tensions that *oriki* reveal, the counter-ideology they occasionally represent, and the subdued voice of the poor and marginalised that are inherent in most."[23]

In Yoruba communities, oriki are often performed by women and addressed to a specific subject.[24] Oriki remake history as present, yet oriki texts are not historical accounts; the items that constitute the chants contain no narrative continuity. Rather, units within the text each have their "own historical moment in which it was composed and to which it alludes."[25] Units, roughly groups of one to four lines, are elliptical and enigmatic, hinting fleetingly at past episodes, or various characteristics. Oriki are partial in both senses, and come with an institutionalized explanatory genre, ìtàn.[26]

I suggest that oriki offer critique that, using Strathern's terms, I point to as decomposing. Events/episodes are decomposed, unsettled, perhaps allowing transitions and transactions between past and present. To illustrate this, I quote the beginning fragment of an excerpt from the text of a performance in honor of Babalọlá, in Elemoso Awo's compound in Okuku where Karin Barber spent many years.

> Èmi náà lọmọ arùkú tí í torí baṣọ
> Ìgbà n mo yẹgọ̀ nígbalẹ̀
> Okú yanhùnyan ó di gbọ́ngan
> Ìdì n mo dirù kalẹ̀ Águrè
> Ewú ọmọ Àlàóọmọ ṣíṣẹ̀ ó wùyá
> Babalọlá ọmọ ṣipá ọmọ́ wu baba
> Irin ẹsẹ̀ ni babaà mi ń fi wu Ìyálóde
> Enígbòórí ọmọ baba Bánlẹ̀bu
> Béniyàn ò bá mi ní tìjòkùn
> Yóò bá mi ní tajígẹ́ẹ̀lú
> Enígbòórí, ni wọn ńki baba Bánlẹ̀bu.
> (I am also the child of one who dons the masqueraders' costume
> When I befit the costume in the sacred grove
> A disruptive corpse was put a stop to

I tied my bundle up at Agure
Ewu, child of Alao, the first steps of the child delight its mother
Babalọla, the child opens its arms, the child delights in its father
The way my child walks delights the Iyalode
Enigboori, child of the father named Banlẹbu [meet me at the dye pits]
If people don't find me in the place where they boil *ijokun* dye
They'll meet me where we go early to pound indigo,
Enigboori, that's how they salute the father called Banlẹbu [meet me at
    the dye pits])[27]

In explaining how the performer, the elderly sister of the Babalọlá being saluted, interpreted these lines, Barber notes that, as she went through her written version of the text line by line, the performer identified the units making up the text. The first unit, lines 1 and 2, was explained as a generalized reference alluding to an identifying characteristic of the lineage, not a specific event. The second unit was line 3, "Okú yanhùnyan ó di gbọ́ngan" (A disruptive corpse was put a stop to).

I quote Barber's account of the old woman's elaboration of this unit below, for it seems to me to be a telling illustration, pointing to the way oriki units decompose events/episodes.

> In [the performer's] hands this brief and obscure formulation unravelled into a complicated story about a real incident in the history of the family. The word *yanhùnyan* is apparently empty, and takes on a definite meaning only when one knows the story—and the only people likely to know it are the members of the small compound to which Babalọla belongs. One of the sons of the family was taken ill in the middle of a masquerading show. They thought he was dead—and to avoid disrupting the show they carried him into the house and hid him. But after the show was over he revived: thus "a disruptive corpse" (*òkú yanhùnyan*) was "put a stop to" (*ó di gbọ́ngan*). The reference is deliberately riddle-like, for not only does it use a word that has no ascertainable meaning until "solved" by a specific application, it also seizes on the most puzzling moment of the whole affair—when the apparent corpse suddenly ceased to be a nuisance by coming alive—and alludes to that alone without explanation.[28]

How does this link to what I understand this book as doing? Let me explain by going back to my stories of Yoruba classrooms in chapter 1. We can understand those stories as decomposing some puzzling episodes in mathematics and science lessons in Yoruba schools, everting a figure of "founded number." It is amusing to think of the image of numbers as sign hooked back to some foundation entity, by analogy to the figure of the "disruptive corpse that was put a stop to."

I can push the analogy even further. My guess is that "the corpse that was put a stop to" featured in an authoritative interpretation of the disturbing event by the knowledge authorities. I guess also that this credited the (non)corpse with a meaning that worked to distribute praise and blame, as I showed the figure of founded number could do for the classroom incidents I told of. The incorporation of the (non)corpse in oriki serves to unsettle that past legislation. In a similar way, everting the figure of founded number unsettles as a decomposing moment.

It is not only their decomposing of events/episodes that has me turning to oriki as inspirational. It is also the ways in which articulation of oriki implicates the past in the present in pointing to futures, and the ways recitation implicates those who recite, that has me turning to them as exemplary critique. It seems to me that what Marilyn Strathern says of the Hagen style decomposing of events—"[it] gives the elicitors of those insides, the decomposers, power as witnesses to their own efforts of elucidation; that the elicitor/witness is in a crucial sense the 'creator' of the image, and his/her presence thus necessary to its appearance"[29]—can equally be said of performers of/listeners to oriki (and perhaps it might also be said of writer/readers of this book). On this reading, performers of oriki, being necessary to the appearance of their own and their antecedent's efforts to elucidate events, hold both a power and a responsibility. One of the performers who features in Karin Barber's book says of herself: "Kùtùkùtù tí mo ríkùn dánà sí" (I who from my earliest days had a mind that could follow many paths). In elucidating this, Barber suggests that "[t]he 'many paths' of social knowledge that Sangowemi alludes to are paths traced by *oriki*. They lead into the thicket of contemporary social relationships and back to influential forebears, 'those whose deeds remain.' . . . They run from the living individual straight back to the moment of origin, when social difference was installed. They allow the past to inhabit the present, perpetually accessible and contiguous."[30]

I feature the work of Strathern and Barber to show the richness and multirootedness of the approach I am struggling for. I do not suggest that the kind of critical analysis I engage in here, and take to be one by which postcolonial times and places might be accomplished, is an exotic form of indigenous analysis. Nor am I suggesting that this is a universal form of critique. The writings of Strathern and Barber, and the forms of critique they identify in places with which they have deep familiarity—the High-lands of Papua New Guinea on the one hand and the town of Okuku in southwestern Nigeria on the other—work for me as a form of inspiration. The resonance interests and inspires me, helping me name what I do in this book.

Inspiration from Science Studies

Much of the critical work in science studies adopting an imaginary of emergence falls under the title of "actor network theory," associated with Michel Callon,[31] Bruno Latour,[32] and John Law,[33] all identifying the work of Michel Serres[34] as a seminal influence. While Law, particularly in his recent work with Annemarie Mol,[35] utilizes what he calls "the perfor-mative move" to consider difference, Callon and Latour tend to focus on how things are "pulled together" in emergence. Donna Haraway's femi-nist critiques of science paint a striking picture of emergent worlds consti-tuted through the work of the biological sciences.[36] Her work shows clearly that telling within an imaginary of emergence cannot attempt an unbiased account. Here, telling is a making explicit, and *why* I want to make some things explicit is intimately tied to *what* I make explicit. Feminist philos-opher Kathryn Pyne Addelson adopts from symbolic interactionism a framing of emergence that she calls "collectivist."[37] Beginning with the insight that foundationism has researchers as privileged judging observers offering moral prescriptions, through consideration of several historical episodes of social work, she shows how outcomes are generated in collec-tive action. At the same time, she considers what a "responsible" telling of these outcomes might be.[38]

I suggest that we might recognize the opening section of Bruno La-tour's classic in the field of science studies, *Science in Action,* as an ana-logue of oriki/ìtàn. The first page of *Science in Action* has three paragraphs, each of nine to ten lines. (I quote a fragment of each.)

Scene 1: On a cold sunny morning in October 1985, John Whittaker en-tered his office in the molecular biology building of the Institute Pasteur in Paris and switched on his *Eclipse MV/8000* computer. A few seconds after loading the special programs he had written, a three-dimensional picture of the DNA double helix flashed onto the screen. . . .

    Scene 2: In 1951 in the Cavendish laboratory at Cambridge, England, the X-ray pictures of crystallised deoxyribonucleic acid were not "nice pictures" on a computer screen. The two young researchers Jim Watson and Francis Crick, had a hard time obtaining them from Maurice Wilkins and Rosalind Franklin in London. . . .

Scene 3: In 1980 in a Data General building on Route 495 in Westborough, Massachusetts, Tom West and his team were still trying to debug a makeshift prototype of a new machine nicknamed *Eagle* . . . [soon to be renamed] *Eclipse MV/8000.* . . . no one was sure at the time if the company manufacturing the chips could deliver them on demand. . . .

Page 2 of the book unravels the little stories, making explicit the links between the solid DNA figure and the configured computer of scene 1, one of which was so ambiguous in scene 2 and the other so doubtful as product in scene 3. By juxtaposing figures everted from three historical moments, Latour has contrived an oriki-like form, duly unpacked in subsequent ìtàn-like text in beginning his book. Pushing this unlikely analogy (perhaps too far), similar to the (non)corpse, which I surmise had a legislated single meaning at the time of its dis/appearance, a meaning that later is both affirmed and unsettled by its appearance in oriki, DNA and Eclipse MV/8000 acquire an accredited singular meaning, which is both further solidified and disrupted in John Whittaker's work at the Pasteur Institute and in Latour's book. Latour's point is that laboratory science is at once both the production of legislated truth *and* the making of generative critique. Latour's disruptive critique is to be understood by analogy to John Whittaker's.

Latour characterizes the brand of science studies he does as "symmetrical anthropology." In contrast with conventional anthropology, symmetrical anthropology has "come home from the tropics and set out to retool itself by occupying a triply symmetrical position. It uses the same terms to explain truths and errors (this is the first principle of symmetry); it studies the production of humans and non-humans simultaneously (this is the principle of generalized symmetry); finally it refrains from making any *a priori* declarations as to what might distinguish Westerners from Others."[39] By dissolving these divides, symmetrical anthropology assumes that all collectives—here or there/"ours" or "theirs"—equally constitute natures-cultures. Or, in my terms, it assumes worlds where, to put things in the form of a slogan, "the material is already and always symbolic, as much as the symbolic is already and always material."[40] When we locate ourselves in this terrain, we can see that what happens in foundationist tellings of the doings of science, as in so many other modern practices, is the making and remaking of the "Internal Great Divide" between the material and the symbolic. This is the duplicitous and coextensive practicing of both purifying and mediating.[41]

Yet, in imagining critique, Latour does not want to jettison everything from the modernity he argues we have never had: "[T]he amalgam [I am looking for] consists in using the premodern categories to conceptualise the hybrids [the emergent material-symbolic relational entities while] retaining the moderns' final outcome of the work of purification . . . as a particular case of the work of mediation [effecting emergent material-symbolic relational entities]."[42]

Toward New Accounting of Generalizing Logic: Numbering, Generalizing, and Certainty

I confront rather similar issues to those my colleagues confront in doing ethnographies of scientific and engineering laboratories, but I must approach them from a different direction. Beginning with what used to be called (in foundationist studies) "the material," "the natural," "the technical," and so on, the task in laboratory studies has been to argue for and focus on the ways the new facts and artifacts generated in laboratories emerge as symbolic and material, as figured and figuring practical routines. Coming from "the social," "the symbolic," "the cultural" side of the foundationist world, the first of my tasks is to show that so-called abstract entities like numbers are as much material as they are symbolic. I need to argue what to many is a ludicrous claim: the materiality of numbers. Pointing to the fingers and toes (digits) lurking in numbers and to the materiality of marks on paper or sound waves that make up words is not enough. Those particular materialities are only a small part of the workings of numbers.

Those coming at this shared endeavor of elaborating the workings of emergent worlds from science and technology studies need to show how certainty in scientific facts is achieved. Locating their stories in the mess of laboratories, this certainty at first seems a very unlikely possibility. In telling the histories of entities, they must show how uncertainties become certainties. Starting off with objects like numbers, with what used to be understood as the abstractions through which generalizing becomes possible, I must show the workings of numbers as material-symbolic objects. For me the first challenge is that most people are unable to imagine numbers, and the other abstract objects of generalizing, as in any way associated with uncertainty. I must work in the opposite direction, as it were, to show the way the certainty that imbues number use clots, and where it is located.

When we opt for emergent worlds constituted in the acting of hetero-geneous material-symbolic entities, the comfortable polarities that figure so prominently in foundationist explanations are dissolved. These polari-ties, which made it so difficult to see that, as Latour puts it, there are no entities that are not simultaneously real (like nature), narrated (like discourse), and collective (like society),[43] are an effect of having worlds known through the propositions of removed, judging observers. Yet as Marilyn Strathern has noted,[44] while the polarizations of object/subject, nature/culture, or found/artificial have crumbled fairly readily in efforts originating in science and technology studies, other polarities have been far more resistant and enduring. The (very productive) polarities of abstract/concrete, particular/general, unique/universal have not been an explicit focus for attention. For me, however, these seemingly invincible polarities loom immediately as an issue. The dissolution of these dichoto-mies becomes a significant focus.

This brings me to the decomposing moment of my critique of founda-tionism, for it is here that these contrasts (abstract/concrete, particular/general, unique/universal) are confronted. Following Strathern, I under-stand this decomposing as "everting a figure." In each of my three decom-posing chapters, this is the figure (or figures) literalized in the previous chapter in securing the differences I elaborate. In chapter 4, it is the figure "recursion"—an odd figure of reoccurring figuring. In chapter 7, it is the figure "features of physical matter." This figure connects the figure of matter/space through time. In chapter 10, it is the figure "matter/space/time." Each of these eversions shows a different aspect of what literalizing achieves in foundationism and how it does so.

Seeing how recursion is literalized in the arguments in chapter 3, we recognize how the "foundness" of the particular instantiations "discov-ered" in foundationist stories is contrived, and how in that process an elaborate ordering exercise is made as a "methodical search." In chapter 7's revelation of "features of physical matter" as literalized, we see how a moral order is contrived in an elaborate framing exercise secured through the literalization. We also recognize that the inevitable conse-quence of this style of argument is to render difference in generalizing logics as outside worlds. In the third decomposing chapter (chapter 10), we see the way foundationist worlds, which proscribe literalizing as a cate-gory mistake, are systematically secured through literalizing the figure matter/space/time. The effect of these decomposing revelations is to dissolve the elaborate boundaries that foundationism demands—the separations of world, representational knowledge, and minds that know worlds through the representations. Instead we find ourselves in a single, complex and confusing emergent domain where those boundaries might be accomplished. My hope is that my new stories, in the chapters that follow the decomposing chapters and that show how those boundaries and others might be accomplished, will provide insight and foresight useful for participants of the times and places those stories are about.

Part Two: Numbering

Chapter Three: A Comparative Study of Yoruba and English Number Systems

In a chapter from my relativist manuscript, I compare Yoruba and English language numbering. I lay out the ways the numeral sequences are generated and used. Next, presenting a relativist account of the origins of number in practice, I explore contrasts that imply significant differences in English and Yoruba numbers. The contrasts appear to be related to the ways number names function grammatically in English and Yoruba sentences.

Chapter Four: Decomposing Displays of Numbers

Beginning with a discussion of the role of numbers in colonizing, I show their collecting and displaying of "the other" as achieved in work of translating and objectifying. It becomes clear that my comparative study of number systems is just such an endeavor, and like all foundationist stories, it deletes the intellectual work of translating and objectifying, insisting on the objects it displays (numerations) as "found."

Next, setting my relativist study of Yoruba and English numbers in the context of previous universalist studies of Yoruba numbers, we see that the differences I showed in chapter 3 were secured through literalizing the figure of recursion. My study explains *away* the disconcertment I told of in chapter 1.

Chapter Five: Toward Telling the Social Lives of Numbers

The chapter opens with a letter from a Yoruba clergyman to a British colonial officer, alleging that inadequate forms of numbering marred the 1921 census in Ibadan. Inspired by this letter, I tell a story of multiple sorts of numbers in early-twentieth-century Ibadan. Next, we see that some of number's remarkable capacities lie in its being a relation of unity/plurality. I follow this by exploring how we might credit the strong sense that many have of number as definitive and deterministic by considering the workings of number as interpellation. This brings us to the point where we can begin to understand the paradoxical notion that numbers are multiple while crediting the sense in which they are singular and definitive. Recognizing possibilities for multiple numbers, we start to see how a politics of difference in number might be understood.

Chapter Three

A Comparative Study of Yoruba and English Number Systems

At the meeting of the Royal Anthropological Institute of Great Britain and Ireland held in London on 9 March 1886, a motion was passed that the paper "Notes on the Numeral System of the Yoruba Nation," prepared by Adolphus Mann, Esq., be taken as read. The report of that meeting, including Mann's paper, published in the journal of the institute, does not enable us to be entirely sure how Mann came to make his "notes." His paper indicates that he had both consulted the "grammars of Yoruba,"[1] available to English readers for some thirty years by then, and made "personal observations" of people using Yoruba numbers in the course of their trading. Whatever the source of his information, Mann is an enthusiastic presenter, giving every sign of taking great delight in Yoruba numbers. Advising his readers that "[a] superficial knowledge, with a slight attempt at praxis, suffices to understand the peculiarities in the arrangement of these numerals . . . [and with this] we light, as it were, on a building, which, when viewed from base to summit is not behind our European systems in regularity and symmetry, while the system surpasses them in the aptitude of interlinking the separate members; it stands to them in the same relation as the profusely ornamented Moorish style stands to the more sober Byzantine," Mann speculates that this wonderful system has its origins in the ways cowries (money) were counted in Yoruba trading.

When a bagful [of cowries] is cast on the floor, the counting person sits
or kneels down beside it, takes 5 and 5 cowries and counts silently, 1, 2,
up to 20, thus 100 are counted off, this is repeated to get a second 100,
these little heaps each of 100 cowries are united, and a next 200 is,
when counted, swept together with the first. Such sums as originate
from counting are a sort of standard money, 20, 100 and then especially
200, and 400 is 4 little heaps of 100 cowries, or 2 each of 200 cowries,
representing to the Yorubas the denominations of the monetary values
of their country as to us 1/2d., 1d., 3d., 6d., 1s., &c.[2]

Adding to the information collected and collated by Mann, a little over
one hundred years later, I too developed an account of Yoruba numbers.
I used information assembled by enthusiastic collectors who came after
Mann[3] and developed further material by consulting grammars and infor-
mants for myself. Like Mann, I was quite enchanted by Yoruba numbers
and turned my delight to painstaking work, developing the text I present
below as part of my project to elaborate a relativist theory of many logics.
I have edited what I wrote some ten years ago,[4] to make more obvious
not only the significant findings of this study but also the assumptions
inherent in its framework of analysis.

Here for the first time was evidence that the grammatical structures
of language might be related to the forms of generalization that we routinely carry out as numbering. In English, number names work like adjectives—they relate to nouns and *qualify*, implying the existence of qualities
in the world. In Yoruba, however, number names work like adverbs. They
relate to the workings of verbs and *modify,* pointing to what we might
call "various forms of manifesting." This implication of a connection between the grammar of language and numbering seems to offer strong
support for the view that there are many different generalizing logics of
numbering.

A Relativist Study of English Language and
Yoruba Language Numbers

Numerals represent the patterns generated in the fundamental practice
of recursive tallying. They constitute an infinite series by having a set of
base elements about which repetition occurs and a set of rules by which
any element can be derived from its predecessor. A numbering system
can be characterized by considering the types of symbols in which the

numeral progression is expressed, the base around which the progression is organized, and the rules for generating successive elements of the progression from their predecessors.

Since numbers are always used in social situations, expanding the inquiry from a narrow focus on numeral systems as such to include some analysis of their use is also important. Considering the ways elements of a number system are used in reckoning operations not only recognizes that numbers work in social situations, but might also tell us something of the origins of numbers. Similarly, since numbers are inevitably part of talk, the grammar associated with inserting numerals into language can tell us something about how numbers are conceived of in a community.

English Language Numeration

Contemporary numeral systems associated with Indo-European languages have ten as the base of their numeral system; in other words, ten is the point in the series that marks the end of the basic set of numerals.[5] As each ten is reached, the basic series is started again, each time recording in the numeral how many tens have been passed by. The rule by which new elements are devised is addition of single units and units comprising multiples of ten.

Number names used in contemporary English have clear links with the linguistic numerals of other Indo-European languages. Etymologically, the basic linguistic numerals can be considered as falling into three groups: (1) one, two, three, four, five, six, seven, eight, nine, ten, hundred; (2) thousand, million; and (3) zero. Numerals in the first group are contemporary English language forms of ancient numerals. They resemble number names in an ancient language that first made its appearance in recorded history some 3,500 years ago as the language of a group that invaded the fringe of Mesopotamia.[6] These first ten numerals provide the base for the synthesis of the first ninety-nine numerals. After ten is reached, the series of basic numerals starts again; this time each member name of the series is joined to the word "ten" in some form. Thus there is eleven, deriving through the Old Teutonic *ainlifun* (one remaining after ten) from the Aryan verb *leip, lieq* (to leave or remain). Similarly, there is twelve, *twen* + *liban* (two remaining after ten). This one-by-one addition continues through thirteen (three + ten) to nineteen (nine + ten). After nineteen, the second decimal point is reached, hence twenty, *twen* (two) + *tig* (decade, from Old Norse).[7]

In English usage, the series of numerals continues in this way: adding ones and tens in a regular linear progression, with each numeral having only one possible successor. At the ninth decimal point, a new basic numeral is introduced, an important numeral naming the square of the base (ten tens): hundred. This name too derives from the ancient Indo-European language. In Sanskrit, it is *kmto-m*, in Latin *centum*, and in Old Teutonic *hunt* or *hund*. The "-red" of hundred derives from an Old Teutonic word meaning "to tell of, or record." In other words, at this point ten tens *(hund)* have been recorded.

The orderly one-by-one additive progression continues with no new basic names being introduced until ten hundreds (another important point in a decimal-based progression) have been reached. Ten hundreds has the name "thousand" in English. "Thousand," like "million," became stabilized relatively recently in the history of this numeral system. In the Indo-European languages, the names for 1,000 and 1,000,000 do not have common origins. The English word "thousand" derives from *ousundi,* an Old English word, an indefinite term for a great multitude that seems to come from the Indo-European word *tus* (a multitude or large force of men). "Million" in English derives from the same root as *mille* (1,000) in French, a Greek word for a great multitude. "Zero" too is of quite recent origin, deriving from the Arabic word *cifr* (cipher). This is the dot used to indicate an empty column in reckoning operations with graphic numerals.

In contemporary English-speaking communities, both linguistic and graphic numerals are commonly used. Numerals are spoken words like "one," "two," "three." They are written words (one, two, three), and they are graphic symbols like Roman graphic numerals (I, II, III, . . . X). Graphic numerals also occur in the form 1, 2, 3, . . . 10. The last is a system of numerals that graphically presents the base-ten number system. It arranges digits in columns that each have a specific base-ten value. This relatively recent addition to the numeral repertoire in Europe[8] was taken over from Sanskrit forms of recording numbers, the practice reaching Europe via their adoption in Muslim cultures.

A second group of numerals in English names, not the sum that has been reached at any point in the number sequence, but specifically the position in that sequence. These are the numerals first, second, third, fourth, fifth, and so on, often called the ordinal numbers. These are represented graphically by combining the cardinal graphic numeral with the ending of the linguistic ordinal numeral, as in 1st, 2nd, 3rd, 4th, 5th.

The terms "cardinal" and "ordinal" are themselves interesting. As names of groups of numerals, they arose relatively recently, after Europeans started to write commentaries on number systems. It was thought that the group named by one, two, three, and so on was more important than the group named by first, second, third, and so on. "Cardinal" has its origin in the Latin word *cardinis* (a major pivot or hinge), while the term "ordinal" derives from the Latin *ordo, ordinis* (sorting into ranks, rows, or sequences). The sense of linear spatial ordering implied by the name "ordinal" is already present in the numerals of this sequence. "Second" derives from *secundum*, gerund of the verb *sequi* (to follow in the sense of along a spatial extension), and the "-th" of fourth, fifth, and so on derives from the Greek suffix *-tos*, used to denote attributes that have degrees of extension, as in the English words "width" and "length."

Yoruba Language Numeration

Yoruba numerals first were written down as words—a series of number names—as the numeration system of which they are part was "collected" by European anthropologists, missionaries, and linguists during the nineteenth and twentieth centuries. Up to that point, the numerals had existed solely as spoken series of names within the actual processes of counting and measuring.

The Yoruba series of number names generates primarily around a base of twenty. Utilizing a secondary base of ten, and then a further subsidiary base of five, integers emerge.[9] There are fifteen basic numerals from which the infinite series is derived: *òkan (ení)*, one; *èjì*, two; *èta*, three; *èrin*, four; *àrùún*, five; *èfà*, six; *èje*, seven; *èjo*, eight; *èsán*, nine; *èwá*, ten; *ogún*, twenty; *ogbòn*, thirty; *igba*, two hundred; *irínwó*, four hundred; *òké*, twenty thousand. The core process in the working of the system is progression from one vigesimal (twenty) point to the next. Multiplication generates multiples of *ogún* (twenty), or to be more exact, "multiple placing out," as the Yoruba verb embedded in the numeral has it. About the vigesimals, intermediate numerals are generated using tens and fives.

Assuming vigesimal points at twenty, forty, sixty, and so on, we can understand the generation of Yoruba numerals in the following way:

> The first four numerals of each vigesimal are generated through addition of ones, say 40 plus 1 (41), 40 plus 2 (42), and so on, a process that is fairly familiar to base-ten users.
>
> After 44 we "leap" to 60 take away 10 take away 5 (60 − 10 − 5) to generate 45; 46 is (60 − 10 − 4), and so on to 49, progressively taking away one less at each step.

Fifty is 60 take away 10 (60 − 10), 51 is 60 take away 10 add 1 (60 − 10 + 1). This continues up to 54.

Fifty-five is 60 take away 5 (60 − 5), 56 is (60 − 4), and so on, progressively taking away one less at each step to 59.

I have expanded this way of presenting the Yoruba numeration system in table 3.1, using the base-ten graphic symbols to illustrate the pattern implicit in each Yoruba number name.[10]

How can we appreciate the complex architecture of this system? Understanding that verbs are essential parts of deriving Yoruba numerals is a beginning. To go further and see the patterns involved in Yoruba numeral generation, we need to know something of the conventions of vowel harmony and elision in Yoruba. These determine the particular ways the basic set of numeral names combine with various verbs.

Table 3.1. Yoruba Language Cardinal Numerals,
Showing Their Derivation

Modern Graphic Numeral	Yoruba Cardinal Name	Derivation Implied in the Name
1	kan	1
2	méjì	2
3	méta	3
4	mérin	4
5	márùún	5
6	méfà	6
7	méje	7
8	méjo	8
9	mésán	9
10	méwá	10
11	mókònlàá	(+1 + 10)
12	méjìlàá	(+2 + 10)
13	métàlàá	(+3 + 10)
14	mérìnlàá	(+4 + 10)
15	méèédogún	(−5 + 20)
16	mérìndínlógún	(−4 + 20)
17	métàdínlógún	(−3 + 20)
18	méjìdínlógún	(−2 + 20)
19	mókòndínlógún	(−1 + 20)
20	ogún	(20)
21	mókònlélógún	(+1 + 20)
22	méjìlélógún	(+2 + 20)
23	métàlélógún	(+3 + 20)
24	mérìnlélógún	(+4 + 20)
25	méèédógbòn	(−5 + 30)

Table 3.1. (*Continued*)

Modern Graphic Numeral	Yoruba Cardinal Name	Derivation Implied in the Name
26	mẹ́rìndínlógbọ̀n	$(-4 + 30)$
27	mẹ́tàdínlógbọ̀n	$(-3 + 30)$
28	méjìdínlógbọ̀n	$(-2 + 30)$
29	mọ́kọ̀ndínlógbọ̀n	$(-1 + 30)$
30	ọgbọ̀n	(30)
31	mọ́kọ̀nlélógbọ́n	$(+1 + 30)$
32	méjìlélógbọ́n	$(+2 + 30)$
33	mẹ́tàlélógbọ́n	$(+3 + 30)$
34	mẹ́rìnlélógbọ́n	$(+4 + 30)$
35	marùúdínlógójì	$(-5 + (20 \times 2))$
36	mẹ́rìndínlógójì	$(-4 + (20 \times 2))$
37	mẹ́tàdínlógójì	$(-3 + (20 \times 2))$
38	méjìdínlógójì	$(-2 + (20 \times 2))$
39	mọ́kòndínlógójì	$(-1 + (20 \times 2))$
40	ogójì	(20×2)
41	mọ́kọ̀nlógójì	$(+1 + (20 \times 2))$
42	méjìlógójì	$(+2 + (20 \times 2))$
43	mẹ́tàlógójì	$(+3 + (20 \times 2))$
44	mẹ́rìnlógójì	$(+4 + (20 \times 2))$
45	márùúndínláàádọ́ta	$(-5 - 10 + (20 \times 3))$
46	mẹ́rìndínláàádọ́ta	$(-4 - 10 + (20 \times 3))$
47	mẹ́tàdínláàádọ́ta	$(-3 - 10 + (20 \times 3))$
48	méjìdínláàádọ́ta	$(-2 - 10 + (20 \times 3))$
49	mọ́kọ̀ndínláàádọ́ta	$(-1 - 10 + (20 \times 3))$
50	àádọ́ta	$(-10 + (20 \times 3))$
51	mọ́kọ̀nléláàádọ́ta	$(+1 - 10 + (20 \times 3))$
52	méjìléláàádọ́ta	$(+2 - 10 + (20 \times 3))$
53	mẹ́tàléláàádọ́ta	$(+3 - 10 + (20 \times 3))$
54	mẹ́rìnléláàádọ́ta	$(+4 - 10 + (20 \times 3))$
55	márùúndínlógọ́ta	$(-5 + (20 \times 3))$
56	mẹ́rìndínlógọ́ta	$(-4 + (20 \times 3))$
57	mẹ́tàndínlógọ́ta	$(-3 + (20 \times 3))$
58	méjìdínlógọ́ta	$(-2 + (20 \times 3))$
59	mọ́kọ̀ndínlógọ́ta	$(-1 + (20 \times 3))$
60	ọgọ́ta	(20×3)
61	mọ́kọ̀nélógọ́ta	$(+1 + (20 \times 3))$
62	méjìlélógọ́ta	$(+2 + (20 \times 3))$
63	mẹ́tàlélógọ́ta	$(+3 + (20 \times 3))$
64	mẹ́rìnélógọ́ta	$(+4 + (20 \times 3))$
65	márùúndínláàádọ́rin	$(-5 - 10 + (20 \times 4))$
66	mẹ́rìndínláàádọ́rin	$(-4 - 10 + (20 \times 4))$
67	mẹ́tàdínláàádọ́rin	$(-3 - 10 + (20 \times 4))$
68	méjìdínláàádọ́rin	$(-2 - 10 + (20 \times 4))$
69	mọ́kọ̀ndínláàádọ́rin	$(-1 - 10 + (20 \times 4))$

Table 3.1. (*Continued*)

Modern Graphic Numeral	Yoruba Cardinal Name	Derivation Implied in the Name
70	àádọ́rin	$(-10 + (20 \times 4))$
71	mọ́kọ̀nlélàààdọ́rin	$(+1 - 10 + (20 \times 4))$
72	méjìlélàààdọ́rin	$(+2 - 10 + (20 \times 4))$
73	mẹ́tàlélàààdọ́rin	$(+3 - 10 + (20 \times 4))$
74	mẹ́rìnlélàààdọ́rin	$(+4 - 10 + (20 \times 4))$
75	márùúndínlọ́gọ́rin	$(-5 + (20 \times 4))$
76	mẹ́rìndínlọ́gọ́rin	$(-4 + (20 \times 4))$
77	mẹ́tàdínlọ́gọ́rin	$(-3 + (20 \times 4))$
78	méjìdínlọ́gọ́rin	$(-2 + (20 \times 4))$
79	mọ́kọ̀ndínlọ́gọ́rin	$(-1 + (20 \times 4))$
80	ọgọ́rin	(20×4)
90	àádọ́ràán	$(-10 + (20 \times 5))$
100	ogọ́rùún	(20×5)
110	àádọ́fà	$(-10 + (20 \times 6))$
120	ọgọ́fà	(20×6)
130	àádóje	$(-10 + (20 \times 7))$
140	ogóje	(20×7)
150	àádọ́jọ	$(-10 + (20 \times 8))$
160	ọgọ́jọ	(20×8)
170	àádọ́san	$(-10 + (20 \times 9))$
180	ọgọ́sọ̀n	(20×9)
190	igba ódin mẹ́wàá	$(-10 + 200)$
200	igba	(200)
300	ọ̀ọ́dúnrún (ọ̀ọ́dún)	(300)
400	irínwó	(400)
500	ẹ̀ẹ́dẹ́gbẹ̀ta	$(-100 + (200 \times 3))$
600	ẹgbẹ̀ta	(200×3)
700	ẹ̀ẹ́dẹ̀gbẹ̀rin	$(-100 + (200 \times 4))$
800	ẹgbẹ̀rin	(200×4)
900	ẹ̀ẹ́dẹ́gbẹ̀rùún	$(-100 + (200 \times 5))$
1,000	ẹgbẹ̀rùún	(200×5)
1,100	ẹ̀ẹ́dẹ́gbẹ̀fà	$(-100 + (200 \times 6))$
1,200	ẹgbẹ̀fà	(200×6)
1,300	ẹ̀ẹ́dẹ́gbẹ̀je	$(-100 + (200 \times 7))$
1,400	egbèje	(200×7)
1,500	ẹ̀ẹ́dẹ́gbẹ̀jọ	$(-100 + (200 \times 8))$
1,600	ẹgbẹ̀jọ	(200×8)
1,700	ẹ̀ẹ́dẹ́gbẹ̀sán	$(-100 + (200 \times 9))$
1,800	ẹgbẹ̀sán	(-200×9)
1,900	ẹ̀ẹ́dẹ́gbẹ̀wàá	$(-100 + (200 \times 10))$
2,000	ẹgbẹ̀wàá	(200×10)

Source: Derived from R. C. Abraham, *Dictionary of Modern Yoruba* (London: Hodder and Stoughton, 1962).

For example, the verb *nọ̀n*, as in *ó nọ̀n* (the state of being placed out or arranged), is involved in formulating numerals. When used with *ogún* (twenty), *ó nọ̀n* shortens to *ọ̀*. Thus in sixty (twenty placed three times), the numeral is *ogún-ọ̀-ẹ̀ta*, which becomes *ọgọ́ta* (sixty) by elision and vowel harmony. In creating multiples of *ogún*, the form of the numerals often changes with obligatory elision and vowel harmony conventions, so that the etymological origins of numerals are not obvious to those unfamiliar with the conventions. Vowel-harmony conventions acknowledge an affinity between *e* and *o*, and *ẹ* and *ọ*. For example, in the two cases where the element multiplying *ogún* (twenty) begins with *e*, like *èjì* (two) and *èje* (seven), the names of multiples of *ogún* begin with *o*, as in *ogójì* (twenty placed two times—forty) and *ogóje* (twenty placed seven times—one hundred and forty). Where the initial vowel of the multiplicand is *ẹ*, the vowel at the beginning of the vigesimal is *ọ*, as in *ọgọ́ta* (twenty placed three times—sixty) and *ọgọ́rin* (twenty placed four times—eighty).

After multiplication of twenties, subtraction is the next most important process in Yoruba numeral formation. The involvement of subtraction is indicated by the inclusion of the verb *ó dín* (it reduces) in various forms. Subtraction is used for deriving numerals from *àrùún* (five) to *ọ̀kan* (one) below a vigesimal or decimal point. Two examples are *mẹ́rìndínlógún* (twenty it reduces four—sixteen) and *mọ́kọ̀ndínlógún* (twenty it reduces one—nineteen). *Ó dín* (subtraction) is used for deriving odd tens from the next highest vigesimal, as in *igba ódín mẹ́wàá* (two hundred it reduces ten—one hundred and ninety). This is also used for deriving odd hundreds from even hundreds, odd thousands from their even counterparts, and odd tens of thousands in the same way (see table 3.1).

Addition is of relatively minor importance in the system. It is confined to involvement with the first four numerals of each decade. The verb *ó lé* (it adds) is used in various forms. For example, numerals eleven through fourteen incorporate the element *làá*, a contraction of *ó lé ẹ̀wàá*, as in *méjìlàá* (twelve). The verb is also incorporated (in a different form) in *mọ́kọ̀nlélógún* (twenty-one).

Putting the step-by-step operations with twenty, ten, and five together with the use of the three verbs *ó nọ̀n*, *ó dín*, and *ó lé*, and using the rules of vowel harmony and elision, it can be seen that there are four basic patterns in numeral formation. I illustrate these four patterns in table 3.2.

While frequently used lower numerals have accepted forms of derivation, multiple versions of larger numbers are possible. Table 3.3 shows seven different numerals for the English language numeral nineteen thou-

Table 3.2. Summary of the Four Patterns Used to Form Cardinal
          Number Names in Yoruba Language

---

1. Forty-one: *mọ́kọ̀nlógójì* (1 + (20 × 2))
   m_́: mode grouped
   ọ̀kọ̀n: mode one
   l_́: elision of *ó lé* (it adds)
   ogójì: elision of *ogún ó nọ̀n èjì* (twenty placed two ways, forty)
2. Forty-five: *márùúndínláàádọ́ta* (−5 − 10 + (20 × 3))
   m_́: mode grouped
   àrùún: mode five
   dín: elision of *ó dín* (it reduces)
   láàád_́: elision of *ó l'ẹ̀wá ó dín* (add ten it diminishes)
   ọ́ta: elision of *onọ̀n ogún ẹ̀ta* (twenty placed three ways, sixty)
3. Fifty-one: *mọ́kọ̀nléláàádọ́ta* (+1 − 10 + (20 × 3))
   m_́: mode grouped
   ọ̀kọn: mode one
   lé: elision of *ó lé* (it adds)
   láàád_́: elision of *ó l'ẹ̀wá ó dín* (add ten it diminishes)
   ọ́ta: elision of *onọ̀n ogún ẹ̀ta* (twenty placed three ways, sixty)
4. Fifty-six: *mẹ́rìndínlógọ́ta* (−4 + (20 × 3))
   m_́: mode grouped
   ẹ́rìn: mode four
   dínl_́: elision of *ó dín* (it reduces)
   ọ́gọ́ta: elision of *onọ̀n ogún ẹ̀ta* (twenty placed three ways, sixty)

Source: Derived from R. C. Abraham, *Dictionary of Modern Yoruba* (London: Hodder and Stoughton, 1962).

sand, six hundred and sixty-nine,[11] and there are still more ways of arriving at satisfactory numeral forms of 19,669. To generate large numerals, one must first be able to render the number as factors, using twenty, ten, and five. Familiarity with factors is what matters most when generating a number name; multiple forms of any one position in the series are possible because there are multiple ways of combining the bases to achieve the same result.

The particular numeral used in any context depends primarily upon the arithmetic predilections of the numerator. Some numeral forms are regarded as more elegant than others; thus the performance of numerators can be judged on the basis of the elegance of their numerals. In large-scale reckoning, operations on each of the bases is dealt with in turn, starting with the twenties and carrying as necessary into the tens and fives. Reckoning with large numerals requires profound familiarity with the system, and in former times this familiarity was fostered during the long apprenticeship that was a necessary part of becoming a numerator in Yorubaland.[12]

Several distinct sets of numerals are found in Yoruba life. Merely naming them here, I will consider these sets and their derivation in more detail later in this chapter. The relation between these derived sets and the primary numeral set can be understood as somewhat analogous to the contrast between cardinal and ordinal numbers in English. One set of Yoruba language numerals is known as the multiplicity set. Here the primary numeral name is prefixed with *m* and a high tone. The most likely verb is *mú*,[13] related to the present-day *mún* (to take or pick up several things in a group or as one[14]). This elision results in a derived form of numerals (one to ten) as follows: *kan, méjì, méta, mérin, márùún, méfà, méje, méjo, mésàán, méwá.*

Another set is often referred to in English as the ordinal numerals,[15] although perhaps "position numerals" is a more precise description, since they are used to indicate position in a sequence. In this set of derived numerals, the verb *kó* (to collect items severally) seems to be involved.[16] This set seems to be prefixed with *ì* + *k'*, where *ì* stands in for an introducer and *k'* stands in for the verb *kó*. However, only in the northern dialects of Yoruba is the elision completely regular. Except in these northern dialects, the high tone is not assigned to the initial vowel of the original numeral. The most common version of the position numerals (first

Table 3.3. Seven Ways of Deriving the Number 19,669

1. òké kan ó dín erinwó ó lé okaàn dínláàádórin
$(((20{,}000 \times 1) - 400) + (-1 - 10 + (20 \times 4))) = 19{,}669$
2. òké kan ó dín òódúnrún ó dín ókànlélóbón
$(((20{,}000 \times 1) - 300) - (1 + 30)) = 19{,}669$
3. èédégbàáwàà ó lé egbèta ó lé ókàndínláàádórin
$((20{,}000 - 1{,}000) + (200 \times 3) + (-1 - 10 + (20 \times 4))) = 19{,}669$
4. èédégbàáwàà ó lé èédègberin ó dín ókànlélógbón
$((20{,}000 - 1{,}000) + (-100 + (200 \times 4)) - (1 + 30)) = 19{,}669$
5. èédégbàáwàà ó lé ótalélegbèta ó lé mésán
$((20{,}000 - 1{,}000) + ((20 \times 3) + (200 \times 3)) + 9) = 19{,}669$
6. èédégbàáwàà ó lé òrinlélegbèta ó dín òkànlàá
$((20{,}000 - 1{,}000) + (-(20 \times 4) + (200 \times 3)) - (1 + 10)) = 19{,}669$
7. òké kan ó dín ótadínírinwó ó lé mésán
$((20{,}000 \times 1) - (-(20 \times 3) + 400) + 9) = 19{,}669$

Source: S. A. Ekundayo, "Vigesimal Numeral Derivational Morphology: Yoruba Grammatical Competence Epitomized," paper presented in the Linguistics Department, University of Ife, 1975.

to tenth) is *ìkíínní (àkọ́kọ́), ìkejì, ìkẹta, ìkẹrin, ìkarùún, ìkẹfà, ìkeje, ìkẹjọ, ìkẹsàn, ìkẹwàá.*

The third modified form of the Yoruba language numerals is the form used specifically in association with currency. In this case, the noun for money, *owó*, is incorporated into the numeral. With elision this elongates, or duplicates, the first vowel of the numeral and imposes a high tone. The first five numerals in this set are *ọkan, ééji, ééta, éérin, árùún.* With the primary numerals *ogún, ogbọn,* and *igba,* the word *owó* is suffixed rather than prefixed, resulting in the numerals *okòó* (twenty units of currency), *ogbọ̀nòó* (thirty units of currency), and *igbiyó* (thirty units of currency).[17]

Origins of Numerations: Zero and Graphic Numerals

Using Yoruba number names was formerly closely tied to performing the actions of counting and reckoning. There are no indigenous graphic numeral forms, and there is no parallel to zero in the Yoruba system. My contention is that these two features are related, and, given the nature of Yoruba numeration, neither feature should be expected. A graphic numeration and the associated invention of zero is likely to arise in single-base numeration systems where mental calculation is difficult, so that the "absence" of both these features in Yoruba numeration reflects the astonishing calculating capacity of the multibase Yoruba numeral system. Having a twenty, ten, five multibase system makes available a plethora of factors from which any number might be devised (as we saw in table 3.3). This leads to an ease of calculation that is difficult for those of us conditioned to a base-ten system to imagine.

At first sight, the absence of zero seems crippling in a numeral system. Many of us are so used to the change between ninety-nine and one hundred being the change between 99 and 100 that we think we need zero to make the change. We do not. The words alone mark the fact that the series has leaped to a new base-ten level. Thinking we need zero here, we are forgetting about the ways words perform, and thinking only that written symbols do the work of a numeral system.

The invention of zero in the Indo-European tradition is considered by many historians of number to have been an event of the utmost importance, and not only because it made negative numbers possible. "[D]iscovery of the zero marks the decisive stage in a process of development without which we cannot imagine the progress of modern mathematics, science, and technology. The zero freed human intelligence from the counting board that had held it prisoner for thousands of years, eliminated all ambiguity in the written expression of numbers, revolutionised the art of reckoning, and made it accessible to everyone."[18] Yet most agree that zero was generated in the confusing mess of actual practice. It seems that zero emerged from a solution to a problem that occurs when base-ten levels are graphed as columns in a two-dimensional array (as 1, 10, 100, and so on). In recording the system of base-ten numerals as graphs, a problem is encountered that does not occur when numerals are words. Graphic numerals that arrange digits in columns that have differing base-ten values respond to reaching the decimal in each level (9, 99, 999, and so on) by placing a 1 in a new column created to the left of the column under consideration. To be workable as a reckoning system, the space in the original right-hand column must also be occupied—otherwise the reckoner can easily make mistakes. It is essential to indicate that the numeral is a double-column or triple-column numeral, and so on.

In the Hindu world where the modern graphic numeral system developed, it seems that numerators came to represent this emptiness in right-hand columns of figures with a commonplace cipher 0—probably just an enlarged dot—which helped prevent muddles. The Arabic word *cifr* became *zero* in Europe when the system of graphic numerals was adopted by Europeans in the eleventh century. *Zero,* the name of a graphic symbol, gradually came to mean "nothing" or "naught." It came to be a number. The naming of a cipher thus eventually resulted in the modern number sequence being given a new point of beginning.

Creation of the number zero grew out of the use of graphic numerals. Thus it seems reasonable to suppose that the absence of zero in the Yoruba number system goes along with, or is associated with, the absence of graphic numerals. The obvious next inquiry is to ask why the Yoruba number system failed to evolve a graphic form. Or, to put the question another way, why did graphic numerals, and eventually zero, evolve with use of Indo-European number systems? The answer is to be found in comparison of the calculating capacities of a multibase derivation system compared to a single-base system. The base-ten number system turns out to be a very poor mental-calculation tool. Not only does it use a single base and an additive generating process, the base, ten, has only two factors—two and five.

In contrast to a base-ten system, which adds single units in its progression, Yoruba numeration, with its multibase nature, multiple sets of progression rules, and the resulting large number of factors, has a capacity to handle calculation that is far superior to the base-ten numeral scale. This is especially so with large numbers. Working out ways to write down operations, that is, inventing graphic numerals, will be felt as a need only when one is working with a poor calculating tool.

It is hard for those of us who grew up in worlds ordered by the modern base-ten numeration to imagine using the complicated Yoruba numeral scale for reckoning. Yet a Yoruba numerator, with a well-honed memory of factorial relations, would scorn the cumbersome graphic processes that must be adopted to remember where you are when calculating with a base-ten system. For a reckoner skilled in the Yoruba system, writing things down would constitute a significant interference in working the system. Besides, how might such a multibase system be written? In the Yoruba system, the pressure developed in actually working the system pushes toward expert oral performance. The absence of a graphic form is exactly what we would expect of such a superior "calculating machine." Since zero seems to emerge with the pressures of the graphic recording of a clumsy calculating system, the absence of zero is also exactly what we would expect of Yoruba numbers.

Differing Forms of Number-Name Generation

Seeing the modest origins of zero in the performance of calculating, where we must remember where we are in a repatterning exercise, we glimpse the origin of number in the practical methods of tallying. The "origin story" for zero-the-number is suggestive of the processes by which the social practices of repetitious arranging and patterning in tallying come to be formalized recursively as the domain of natural number, which in turn is represented as a recursive system of symbols.

There is general acceptance that numerations derive from patterns that use the human digital sequence—fingers and toes—as a template.[19] We can recognize this bodily template in both the Yoruba and English system of number names, but the template appears in different forms.

We might say that the Yoruba language rules for numeral generation can be taken as modeling the following sequence of specified actions with hands and feet, fingers and toes. Beginning with the finger/toe complement of a person gives the major base *ogún* (twenty). Shifting from one vigesimal to the next codes for "starting a new set of fingers and toes"— literally "placing out a new set of twenty" in Yoruba; for example, sixty (20 × 3), *ogóta*, is twenty placed three times. Specifying *l'ęwa ó dín* (reduce by ten) signifies "hands only," and the additional specification *l'árùún ó dín* (reduce by five) signifies "one hand only." At this point we have $(-5 - 10 + (20 \times 3))$, *márùúndínláàádọ́ta*. Continuing in this vein, the next specification is "reduce the reduction by ones" $(-4 - 10 + (20 \times 3))$, *mẹ́rìndínláàádọ́ta*; and along the fingers and so on to the next hand and along its fingers $(+1 - 10 + (20 \times 3))$, *mọ́kọnléláàádọ́ta*. Next specify "toes (one foot)" $(- 5 + (20 \times 3))$, *márùúndínlógọ́ta*; then "reduce the reduction by ones, working along the toes" to "add the toes of the other foot by ones" $(+4 + (20 \times 3))$, *mẹ́rìnlélógọ́ta*. Notice that the first *ogún* (twenty) is not specified before starting. Tallying works toward the first *ogún*, not back from it. Presumably, the numerator starts with his or her own fingers/toes. Yoruba numeral generation seems to imply a picture of numbers nested within each other as hands and feet are nested in persons, and fingers and toes are nested in hands and feet.

This is very different from the picture we get with the base-ten numerals used in modern quantifying and mathematics, say in English. There ten implies the one-by-one pointing to the ten fingers. Twenty implies two sets of fingers. Twenty-one begins again: thumb of left hand held up; twenty-two: forefinger of left hand, up; twenty-three: tall man of left hand, up; and so on to twenty-six: thumb of right hand, up; and on up to thirty, which is three sets of fingers. What we see in English is a linear passing along the fingers until the end, keeping tally of how many sets, and then doing it again. Each finger/integer is equivalent and related to the one before and the one after it in a linear array.

Interestingly, the contrasting patterns seen in the generation of English and Yoruba language numbers echo a contrast found in accounts of number given by Western mathematicians. Von Neumann and Zermelo both seek to explicate number by identifying the referents of number words, but they have come up with quite different, and contesting, explications. Their enterprise is very different from the one that engages me;[20] nevertheless their differing accounts of number can be useful in helping us see the contrast in the patterning of English language and Yoruba language numeral recursions more clearly.

As an English language speaker, von Neumann's account seems to me to be intuitively correct. Cardinality on his account seems to encode a one-by-one collection of predecessors. Von Neumann has the members of an *n*-membered set paired with the first *n* numbers of the series of numbers. For von Neumann, zero is $\varnothing$, the empty set. The set that contains the empty set as its sole member is $\{\varnothing\}$, one; the successor of this number is $\{\varnothing, \{\varnothing\}\}$, or two; and three is the set of all sets smaller than three $\{\varnothing, \{\varnothing\}, \{\varnothing, \{\varnothing\}\}\}$. In von Neumann's version, the successor of any number is generated by adding the successor of zero, that is one. A number here is the last number of the series reached through one-by-one linear progression.

In contrast, I suggest that a Yoruba speaker would intuitively choose Zermelo's account of number as correct. Here the number n is a single-membered set; the single member of the set is $n - 1$. Zermelo has zero as $\varnothing$, a set with no members. Then 1 has the empty set as its sole member, $\{\varnothing\}$. The set that contains the unit set, 1, as its sole member is 2, $\{\{\varnothing\}\}$. Three is the set that has two as its sole member, $\{\{\{\varnothing\}\}\}$. In this version, each number is totally subsumed by, or nested within, its successor, and any one number has a unified nature. I contend that for a Yoruba speaker, the model of number that would jump out as the intuitively correct account would be this one, for Yoruba language numbers carry the flavor of multiply divided, nested wholes; there is no sense of a linear stretching toward the infinite.

The two formal models agree on the overall structure of number. Each model is demonstrably a recursive progression, although, importantly for many mathematicians, in von Neumann's account (and English language numbering) the set 14 has fourteen members, while for Zermelo (and Yoruba language numbering) 14 has one member only. While this is a significant difference for mathematicians, for ordinary users of number, with no interest in models of cardinality, the difference is irrelevant. For them, that numbers work is all that matters.

Yoruba numbers are generated using the complete set of human digits as a template. The bodily divisions—the distinctions between hands and feet, left and right—seem to be echoed in the structure of the base of Yoruba numeration, and in the rules for deriving numerals. This contrasts with English language numeration, where it seems that only the hands (or feet) are templates, and the ten fingers (or toes) are treated as a set of homogeneous elements taken as linearly related.

What is the origin of this difference? What in the practice of tallying has a Yoruba numerator turning to the template of all his digits, picking up on the separations and the nestings of the four groups of five digits in the process? On the other hand (so to speak), why does the numerator in a community that speaks an Indo-European language like Sanskrit refuse the clearly apparent distinctions between the sections of the human digital complement in devising a template for tallying? What is the difference in the apparently identical project of ordering a collection (of sheep, soldiers, cowrie shells, etc.) in these communities that has the groups

using the same template (the human digital complement) in such contrasting ways?

I suggest that what makes the difference is the languages the numerators and their clients speak. Words too are a material part of the activity
of tallying. Keeping count involves more than fingers/toes and sheep/
cowrie shells. Thus I now consider Yoruba and English number names
as words constituting ordinary speech.

Numerals in Language Use

Yoruba number names are constituted as elided verb phrases, and as
such, they occur in Yoruba sentences in ways that are very different from
the occurrence of number names in English sentences. Unlike the latter,
they are in no way adjectival, ascribing and describing qualities.

The role of verbs in Yoruba number names is central. Earlier I elaborated the fifteen basic numerals of the set of cardinal or "counting" number names in Yoruba, and we saw that the verb phrases *ó nòn, ó dín,* and
ó lé are an integral part of their working (table 3.2). In taking this further,
I begin by looking again at the primary numeral forms (not the cardinal
forms): *òkan (ení),* one; *èjì* , two; *èta,* three; *èrin,* four; *àrùún,* five; *èfà,*
six; *èje,* seven; *èjo,* eight; *èsán,* nine; *èwá,* ten; *ogún,* twenty; *ogbòn,* thirty;
igba, two hundred; *irínwó,* four hundred. These numerals are ancient, and
etymologically their origins are hazy. Nevertheless, through applying the
rules for elision and vowel harmony in Yoruba to tease the elements of
the terms apart, we can still gain etymologically significant insights.

As elisions of introducers and verbs, the names of the counting set of
Yoruba numerals have the form of nominalized verb phrases. Saying that
the numerals are elided verb phrases identifies that grammatically the numerals function as mode or modal nouns. "Mode" here is related to the
modifying function that adverbs have in sentences. Number names are
adverbial in nature, ascribing "a manner of appearing." It seems that Yoruba number names relate to the verbs in the sentence, not the nouns.
This may be difficult for English speakers to understand, so used are we
to number names referring to nouns as they do in English. It seems that
òkan is best translated as "mode one" and *èjì* as "mode two"—they imply
a particular form or arrangement in manifesting.

The mode nouns of the primary numeral set are not used in conjunc-
tion with other nouns. These mode nouns constitute the counting series,
and when we recognize them as mode nouns, we can see that they are

naming a series of appearances. A sequence of modes, or differing forms of manifesting, is named by the sequence of Yoruba numerals in counting.

There is an additional point to be made here about the absence of zero in Yoruba numbering. The absence of zero does not imply that the idea of nothing is absent from Yoruba thought. The notion "naught" or "nothing" in Yoruba is *àìwà* or *àìkanwà*. In these words, the *àì* indicates negation, and *wà* is a verb (to manifest); *kan* indicates "one." Literally, *àìkanwà* means "not as one manifesting" and *àìwà* means "not manifesting." They are not considered to be numerals.[21] Only in very odd circumstances would there be a point in speaking a word (conceivably "àìkanwà") to announce the nonappearance of the matter to be tallied as a beginning of the act of tallying.

This brings me to the issue of well-intentioned attempts to "modernize" the Yoruba numeration system, in part by adding zero. A scheme of reform was developed in 1962 by Robert Armstrong when he was director of the Ibadan Institute of Education, in which "[a]ll subtractive numerals are abolished, and a zero is added at the beginning . . . [so that] any of the operations of arithmetic can be easily expressed verbally in this [proposed] decimal system."[22]

In commenting on my manuscript "Numbers and Things" in 1986, Karin Barber noted that

> the education authorities have already tried to "simplify" the Yoruba numeral system by removing one stage in the process, i.e. the fives up-and-down between the decades. Now they are copying the Indo-European system and adding from one to nine from the previous decade: *méjelógún* instead of *métàdínlógbòn*. At the same time they've simplified the construction of the words: *métàlógún* instead of *métàlélógún*. As they've left the construction of decades intact, I don't know whether the new system will make it easier or just confuse things further. . . . The bridge between the two conceptual systems they were beginning to teach [in Okuku] before I left is only a bridge . . . at a very superficial level.

In considering to what extent reform of Yoruba numeration is needed, and likely both to foster the survival of Yoruba forms and to help contemporary Yoruba children as they work in the modern world, we should remember the ways zero, developing as a trick in graphic recording of numbers, eventually became a number in its own right in a wide variety of cultures. Like any cultural product, number systems evolve. However, we also need to note that the linguistic differences between Yoruba and the Indo-European languages, through which zero diffused so easily, are large, and these differences are embedded in the connotations that the number names carry as they are spoken. There is a need to ensure the continued coherence of the Yoruba system, and when it comes to learning/teaching, there is a need to guard against the possibility of inadvertently introducing cognitive dissonance in inventing a "reformed" numeration system. These imply that "improvements" such as those proposed by Armstrong should be treated with great care.

The sense that Yoruba number names, as mode nouns relating to verbs, point to performance is further developed in the sets of numerals derived from the primary set. The primary numeral form (already a mode noun, an elided phrase with verb and introducer) is further modified when used in quantification statements in Yoruba talk. In the multiplicity set, sometimes called the cardinal set, the primary numeral name is prefixed with *m* and a high tone—the nominalization of a verb phrase that already contains a noun previously derived in the same way. Thus the derived numeral forms in Yoruba seem to imply modes of modes!

The most likely verb involved in generating the multiplicity set is *mú*,[23] an obsolete verb related in meaning to the present day *mún* (to take or pick up several things in a group or as one[24]). This elision results in *kan, méjì, méta, mérin, márùún, méfà, méje, méjo, mésàán, méwà*. The numerals imply the mode of being grouped in addition to the mode of being one, two, three, and so on. Thus one can say, "Ó fún mi ni ókúta mérin," which is conventionally translated as "He gave me four stones." A more literal translation is "He gave me stonematter in the mode of a group in the mode of four." Similarly, we can say, "Ó rí ajá méta" ("He saw three dogs"), which is precisely translated as "He saw dogmatter in the mode of a group in the mode of three." The numeral *òkan* does not accept the prefix *m* + (high tone); instead, the first vowel of the primary numeral *okan* is dropped to form *kan*, which may be used as follows: "Ẹja kan kò tó" (One fish is not enough), literally "Fishmatter in the mode one does not reach." Perhaps the failure of *òkan* to accept combination with the verb *mú* relates to *òkan*'s already being in the mode of a group.

This remarkable double modal form is also found in the position numerals, where the verb involved in the elision is *kọ́*. A literal translation of "Ó gbà ìwé ìkẹta ni" (He took the third book) illustrates the double modal nature of this form of numeral: "He took bookmatter in the mode of collected individual items, in the mode three." *Ìkẹta* is a double mode noun formed by nominalizing a verb phrase that already contains a mode noun, the primary numeral. In contrast to the implication in the multiplicity form, where the individual elements lose their individuated existence, the implication here is that the constituent elements retain their "oneness."

In contrast to the modal nature of Yoruba numerals, in modern English usage "ten" in the sentence "I cooked ten potatoes" is taken as telling about the extent of the quality of numerosity held by that group of potatoes boiling in the pot. Numerosity is a quality or property of the collection of potatoes held to a particular extent—the ten single brownish lumps. Similarly, in "She is five feet tall," the "five" tells about the extent of the quality or property of length in the body of a particular woman. In English, number names work as adjectives; they qualify in a second-order way. Despite this, most English grammars class number names as abstract nouns, following the official line that numbers are naturally occurring abstract entities.

When we see how Yoruba number names are inserted into ordinary language, we recognize that their use in talk implies modes of presenting: particular arranged manifestations. In everyday use, they are modes of modes implying particular further arrangement of previously arranged presentations. Grammatically there is no adjectival element in Yoruba numeral use, implying that those mysterious abstract entities, qualities, or properties, so central in scientific quantifying, do not appear to be part of Yoruba quantifying at all. Whereas Yoruba number names imply a second-order form of modifying, English language numerals are a form of second-order qualifying.

It is this characteristic of working through modes that has an analyst like Hallpike classifying Yoruba numeration as primitive.[25] On his universalist account of quantifying, processes that proceed in ways that fail to evoke the abstract element of qualities are primitive. Similarly, Carnap's positivist account has the evoking of qualities as central in the development of quantitative concepts.[26] Here we see the significance of a relativist account of number like Bloor's.[27] This has social practices of ordering as constituting the foundation for the domain of number, and allows for the possibility of numbers being modes as easily as it accepts the possibility of their being qualities. We begin to recognize the possibility of there being many separate logics of generalizing, each recognizable only as relative to others.

Chapter Four

Decomposing Displays of Numbers

*I*n developing the comparative, relativist study I reported in chapter 3, I spoke to many Yoruba people about Yoruba numbers. I watched how Yoruba and English number names were used in many everyday situations like markets, as well as in classrooms, and I struggled to make conversation in Yoruba, using numbers. In this chapter, I show that the links between my relativist study and the universalist studies that informed it were actually much stronger than I understood at the time I wrote chapter 3. My study of Yoruba numbers was "anticolonial" in intent, yet a close examination of its textual strategies reveals that it was itself contained in, and contained, the very enterprise it set out to unmask. Here I critically examine the methods of my arguments in chapter 3, intending this interrupting as a transitional moment. I recognize now that an unnoticed literalizing secured the difference that I showed between the two numbering systems. Like all foundationist studies, the arguments of chapter 3 depend on literalization, or self-reference.[1] Setting the difference through literalizing has the effect of denying its reality in the sense that it might be the subject of negotiation and managed by variously making connections and separations.

I begin this chapter by considering the role of numbering in colonizing as a way of alerting readers to the role of various scholarly practices of translation in *generating* numerations as objects—the items of the comparative display in chapter 3. My relativist study, like its universalist predecessors, deletes mention of this scholarly work. The convention is that we scholars find the numerations already objectified awaiting our "collection," for particular numerations must be "found" if they are to instantiate the domain of natural number. We say that the numerous contrivances we make in our studies are the methods of our search. Seeing this contrived "finding" enables me to recognize points in the text of chapter 3 where "tallying recursion" is simultaneously a thing *and* its representation.

Collecting, Translating, Objectifying, and Colonizing

Arjun Appadurai begins his exploration of "number in the colonial imagination" with Edward Said's aphoristic summing up of orientalism: "Orientalism is absolutely anatomical and ennumerative; to use its vocabulary is to engage in the particularising and dividing of things oriental into manageable parts."[2]

Appadurai argues that these particularizing and dividing projects that "manage the other" exemplify the colonial work of numbering. He identifies two rather distinct ways in which colonizing is enumerative. The first effects a translating, and the second an objectifying. The distinction is marked by tension between those officials concerned with local variation and on-the-ground accuracy and fairness, and those attending the demands of the bureaucracy, with the former resenting what they perceive as the grand and often unrealistic expectations of higher levels of the bureaucracy who forget about the mess that numbers order. I contend that the study of number systems elaborated in chapter 3 can be seen as exemplary translating and objectifying and that its hiding of this work aligns it with the colonizing project of its predecessors.

The first type of colonial numbering could be labeled cadastral. It aims to achieve the most encyclopedic and general range of application as well as providing a sensitive accounting of taxable resources. This cadastral version of colonial enumeration performs like a theory of the world. Tallying goes along with classifying. In a straightforward way, in the records and files of any colonial office, numbers manage mess and heterogeneity.[3] "Numerical tables, figures and charts allowed the contingency, the sheer narrative clutter of prose descriptions of the colonial landscape, to be domesticated into abstract, precise, complete, and cool idiom of number . . . far removed from the heat of the novel, the light of the camera, and the colonial realism of administrative ethnographies."[4] As a theory of the world, the cadastral moment of colonial numbering effects a translating and an ordering.

The second sort of colonial numbering that Appadurai recognizes features in different places in the work of empire. We might gloss this as imperial number. It is displaying rather than collecting, and this moment effects what has been collected as object. Numbers occupy and fill in tables, charts, and figures that become published documents: reports and booklets. These numbers perform the empire to itself; they are generative of "the Empire." As Appadurai points out, the collective imagination of emergent, nineteenth-century European states held that a powerful state could not survive without making enumeration a central technique of social control. Imperial numbers generate a map by which the empire recognizes and knows itself. In this doing of number, the territory *is* the map.[5]

> The vast ocean of numbers, regarding lands, fields, crops, forests, castes, tribes, and so forth, collected under colonial rule from very early in the nineteenth century, was not a utilitarian enterprise in a simple referential manner. . . . statistics were generated in amounts that far defeated any unified bureaucratic purpose. . . .
>
> . . . numbers gradually became more importantly part of the illusion of bureaucratic control and key to a colonial *imaginaire* in which countable abstractions, both of people and of resources at every imaginable level for every conceivable purpose, created the sense of a controllable indigenous reality.[6]

How does this account of numbers and colonizing relate to my comparative study of Yoruba and English numbering that I elaborated in chapter 3? I suggest that the largely unnoticed translating and objectifying work in my study, which effects its collecting and displaying, is analogous to the colonial work of numbering. It is in my translating and displaying work that the number systems I "found" *became* objects. Duplicitously, I present them as waiting there, ready for collection, and in this my study is aligned with those universalist studies that preceded mine, and with their colonizing project.

Quite obviously, since I am writing in English, I translated Yoruba words into English, struggling to find adequate ways of conveying meanings that are not too clumsy when rendered as English words. In showing the structure of Yoruba numeration, I also translated between the patterns of generating new positions in a twenty, ten, five multibase system and those generated with base ten, a numeration with which most of my readers are familiar. My translating, like that of Adolphus Mann, with whose study I introduced the Yoruba number system, can be understood as a form of standardizing, as finding a form to bring the Yoruba numeral system "home" to display its beauty and its unusual features to my colleagues.

This double translating, between languages and between numerations, is an integral part of achieving a standardized form for display. It can be compared to the drawings of objects collected in earlier colonizing endeavors. As Nicholas Thomas points out,[7] in Captain James Cook's *Voyage around the South Pole* we see standardized graphic depictions of artifacts collected from various islands as the expedition wended its way across the Pacific Ocean. Irrespective of the islands from which they were collected, Maori, Malekula, and Tannese axes, spears, flutes, and fans are transparently rendered as artifacts in the considered arrangement of drawings on the pages of Cook's journal.

In that display, much attention has been paid to scale of presentation and the placement of the diagrammatic images on the page, to create a pleasing, quite standard aesthetic. The collected objects have been translated into the form of a journal, and in this the islands from which the objects were removed, and the people who fashioned the objects, figure hardly at all—reduced to annotations in small print placed just so, as a title. The implements emerge as elegant, with an aura of beauty, although as Nicholas Thomas points out, this standardized rendering as beautiful did not extend to the makers of the tools.[8] Thomas's point is that, in the exchanges of colonialism, the taking, theft, or purchase of objects leads to creative recontextualization, and even reauthorship. Museums, both in the past and in the present, were and are prominent in that "rewriting."

Extending this insight to something like a numeration system allows us to see a side of this recontextualization that can easily be missed when the focus is on actual objects picked up in the arms of sailors, carried on board a ship, sketched, crated, and later unloaded in England to be transported to the British Museum. When we are focusing on a numeration system, the disembedding/reembedding, disentangling/reentangling, and disembodying/reembodying work of the translations in generating standard forms becomes more obvious. We might understand this as a transfer between collectives. Spears, axes, carrying baskets, flutes, fans, and hats have meaning and are participants in collective going-on among the Maori, Tannese, and other men and women who made them. They evoke quite different connections, and participate in quite different ways, when encased in glass-and-wood cabinets, watched over and guarded by uniformed attendants in the British Museum. There is an *objectifying* occurring here as well, which is easily missed because in an experiential way fingers can be enclosed over spears, axes, carrying baskets, and hats as things in both contexts.[9] In their former lives, they were not necessarily objects; they may well have been extensions of human bodies.

Along with Adolphus Mann, I have translated between languages and differently based numerations, and importantly, I have translated in another way, too. As well as translation from Yoruba language to English, and from a twenty, ten, five multibase system to a base ten, we have disentangled the Yoruba numeral system from its embedded and literally embodied way of contributing to the ongoing life of Yoruba trading, and reentangled it in the pages of books and journals. By getting it onto pages and into books in reembodying it, we have literally objectified it; we have rendered it as an object—the Yoruba numeration system—whereas before our interventions there was not an object at all. Before it was made into charts printed on pages in books as paragraphs of descriptive and explanatory English words, Yoruba numerators and ordinary Yoruba people knew their counting in particular ways. The patterns made in spoken words contributed to the life of the markets of Yorubaland, so that it was scarcely possible to separate human body and numeral. Each person had their own way of working the pattern according to their own aesthetic, even though the numeral words, when used, evoked certainty, specificity, and precision.

In being put onto paper and into books, what were spoken words, actions of hands, and feelings about patterns became a thing—the Yoruba numeration system. We understand it now as a something, like a separate material object in space and time with certain properties. As Mann tells us, it is "a building" that may be "viewed from base to summit." As a standard object, the Yoruba numeration system has become a particular example of the "class of all numeration systems." All examples of this class have a history like that of the Yoruba numeration system: they came to life—were born as objects—in the cadastral/imperial work of European colonizing. Particularizing and classifying effects translating and objectifying, which, in being denied, serves the colonizing project.

As an example of an instantiation of the group of entities that constitute the "class of all numeration systems," the Yoruba numeration system is a particular that has certain qualities. As Mann points out, it has "regularity" and "symmetry." Having qualities, the Yoruba numeration system can be compared to other numeration systems, other instances of the "class of all numeration systems." It becomes available for, and in, generalizing.

As Mann opines, "it surpasses our European systems in the aptitude of interlinking" and has "a profusely ornamented Moorish style," compared to the sober Byzantine style of European examples of the "class of all numeration systems." Of course, not only the numeration system is being objectified in Mann's paper. The "Yoruba Nation" of Mann's title is similarly emergent in the cultural work of Yoruba ethnogenesis that was just beginning.[10]

Why Contrive to "Find" Numerations?

In my presentation, I showed a "found" pattern. Yet, in "finding" that pattern, I did not notice that I could only show the pattern, as distinct from feeling it when I uttered words in specific contexts, by working up a systematic translation of one set of patterns into another (the twenty, ten, five multibase recursive patterns into the base ten recursive patterns). I also "found" that the names of the Yoruba counting sequence were mode nouns. I showed this through elaborate and painstaking translations of single Yoruba words into clumsy English phrases. Next, I "found" that numerals in Yoruba sentences are double mode nouns, and I had to show that by even more elaborate and clumsy English phrases.

There is no doubt that Yoruba numerals exist as words spoken in the ongoing collective acting of Yoruba markets, for example, and that as spoken words they weave patterns that people inhabit and that inhabit people. Yet the *object* Yoruba numeration system is born in the work of translation—in the showing. Perhaps it was my (and others') enthusiasm and delight in the patterns that blinded us. The scam seems so obvious. It is quite an accomplishment to seamlessly conflate and elide a set of felt patterns made in spoken words that contribute in the life of the markets of Yorubaland, a performance scarcely separable from the acting human body, with something literally objectified as a chart printed on pages and paragraphs of descriptive and explanatory English words in books, which contributes to the ongoing collective life of the academy and official discursive life.

For years as I puzzled about the moral and political problems that come with a foundationist outlook, I remained convinced that I had accidentally stumbled across, found, all the signs of the existence of the object—Yoruba numeration. I failed to notice that if I had been just going on using the number words, say, in buying provisions for my family in the market, concentrating only on the price of the bowl of rice (insisting loudly that the vendor heap it up), or pile of tomatoes (noting their overripeness in my negotiations), bargaining with the vendors to get a better price, there would have been no resources to make an object with. It is precisely that I brought the object of the Yoruba numeration system "home" to the academy, rather than just the rice and tomatoes home to my family, which provided the materials to generate the object "Yoruba numeration system." Invisibly constituting the object of the Yoruba numeration system so that it might be taken as found, I remade the necessary boundary between particulars and general category.

Making the claim that the historically common social practice of recursive tallying rendered a category that now is variously instantiated in several numerations involves me in displaying the "found" numeration systems of contemporary English and Yoruba as my evidence. This is thoroughly orthodox making of knowledge claims in the modern academy. The knowledge claim can be evaluated. I back up my claim about the nature and origin of the general category, the foundation, by presenting observations of particular instances. A would-be evaluator can question whether the evidence is sufficient and relevant in establishing the proposition concerning the general category. I assemble my evidence in the light of this evaluative regime. It is in my interest to fail to notice the duplicity upon which my knowledge claim rests. If I noticed that the particulars or objects (numerations) I describe as the basis for my authoritative statements have been generated in the endeavor of research, in my translations, there would be no point to it. Making knowledge claims within this style of reasoning would be revealed as self-vindicating activity.[11] As I go on, systematically ignoring evidence of my own participation in generating the objects that constitute my "found" evidence, I regenerate myself as authority who has privileged access both to the domain of particulars and that "other" domain of the general, and my project lines up with those of my universalist predecessors embedded in a certain moral economy. Contriving to "find" my evidence and hiding the contrivance betrayed my relativist project, just as it betrayed my universalist predecessors.

Displaying a Hierarchy of Human Languages and Races

Why were Adolphus Mann and his colleagues in the Royal Anthropological Institute of Great Britain and Ireland interested in Yoruba number names? Is it not a little puzzling that a century ago, perhaps in Lagos, Mann squats down uncomfortably with his notebook, sweating in a crowded room in a market—the cloth section perhaps, where richly woven fabric strips for making matching outfits for vast, entire families change hands? Possibly causing much hilarity and not a little embarrassment to himself and others, he makes careful notes on the actions of the numerator who has been called in to publicly count the cowries to be handed over. What motivates Mann to weather this discomfort? Back in London, what moves him to transcribe and expand his notes, and put together a written presentation to be given to his friends of the Royal Anthropological Institute of Great Britain and Ireland on 9 March 1886? What has the recipients reading it, and leads the Royal Anthropological Institute of Great Britain and Ireland to officially "take the presentation as read"?

In 1886, Adolphus Mann's work was part of a vast cataloguing endeavor to generate an "objective scale of civilization" aimed at understanding how "civilized European man" had attained his eminence, so that other lesser sections of humanity might be helped. This vast cataloguing enterprise understood itself as attending to what was seen then as the central question of scientific endeavor: turning the experimental project to developing an account "of man and his place in nature."[12]

What I mean to draw attention to here is that the endeavor creating the arena in which the relatively newborn object, the Yoruba numeration system, came to life, was one that understood itself as a project of the highest moral standing. What seems despicable to us now, an intellectual project to develop a scale of humanity, was not recognizable as such at the time of its generation. To acknowledge the good intentions is of course not to condone the horrific treatment that many Africans suffered at that time at the hands of Europeans, and that the work on number systems to some extent justifies.

Adolphus Mann helpfully notes in beginning his presentation:

[O]f late the nations and languages of West Africa have largely occupied the attention of the learned linguists of Europe, and grammars and vocabularies are being published in considerable number. . . . the classification of four or five hundred languages has been advanced to such an extent as could not some years ago have been expected. Perhaps the following notes of the numeral system of the Yoruba nation may . . . be of some use in investigating the nature of the mind that can form such an unusual, yet regular structure.[13]

The endeavor was not confined to Europe. A few years earlier, one C. H. Toy, professor of Jewish studies at the Southern Baptist Theological Seminary, Louisville, Kentucky, had addressed twenty-five other philologists at the tenth annual session of the American Philological Association, in the Opera Hall of the Grand Union Hotel in Saratoga, New York, on Tuesday evening, 9 July 1878. His talk, entitled "The Yoruban Language," was, we are told in the record of the transactions of that meeting, remarked upon by Professor S. S. Haldeman, University of Pennsylvania, Philadelphia.[14] Some time later, his paper, with a section focusing specifically on Yoruba numbers, was published in the record of the transactions of the association as "The Yoruban Language." Along with this paper, we find others with such titles as "Contributions to the History of the Articular Infinitive," "Influence of Accent in Latin Dactylic Hexameters," and "Elision, Especially in Greek." Toy explains his motivations in this way.

> The main body of African languages . . . falls into three groups: The Hottentot in the south, the Bantu, occupying the whole center . . . and the Negro lying in Senegambia and Soudan, the last of which has as yet received little attention, while the structure of the others has been carefully studied and satisfactorily exhibited. On the Guinea coast, however, there is found a group of dialects wholly different in vocabulary and structure from all these and offering interesting linguistic features. . . . the most important member [of this group] is the Yoruban which is spoken by a partially civilized population of about two million people.[15]

Mann and Toy are interested in the characteristics of language systems, considering each of them as an instance of "the class of systems of representing the real world." They assume that each language has particular characteristic ways of representing the real world, and they are interested in "the evolution" of languages as an indicator of the evolution of races. Their displaying, classifying, and cataloguing of languages is a way of exhibiting a hierarchy to display the various stages in the progressive evolution of language. The Classical languages of Greek and Latin serve as reference points in this enterprise, and necessarily, languages are constructed as standardized objects.[16]

Number systems are of particular interest in this endeavor because numbers as abstract real objects point to a degree of civilization. The type of number system that a language incorporates is an important sign of its "degree of development." It is a convenient basis, a significant indicator, we would say these days, for classifying languages on a scale of adequacy in giving access to the "real" world. Sure in the knowledge that the inferior systems of "others" are various stages in the development of "our" advanced modern representations of mathematical universals, these books sought, and still seek,[17] empirical, as distinct from philosophical, answers to questions about the origins and development of mathematical objects. Most treat the base-ten numeration system of modern mathematics as a perfected tool, whose perfection has been achieved through the dual determinants of a real world of number and a superior rational mind particularly attuned to that real, though abstract, domain of number. Less advanced systems are those developed by "other" cultures using inferior languages, and minds less well fitted to recognize that ideal domain of number.

In this literature, there is continuing disagreement over whether or not all peoples have at least some degree of natural affinity for the abstract domain of natural number,[18] and over whether the domain is accessible to nonhuman animals.

Levi Conant's 1896 book, *The Number Concept: Its Origins and Development,* is a rather extreme example of the genre. Conant is sure of the extraordinary capacity of "modern man" to apprehend that domain, and of the perfection of the modern numeration system.[19] Animals, he thinks, might have a hazy awareness of number, as might some "others." He approvingly quotes

> an amusing and suggestive remark in Mr Galton's interesting *Narrative of an Explorer in Tropical South Africa.* After describing the Demara's weakness in calculation he says: "once while I watched a Demara floundering hopelessly in a calculation on one side of me I observed 'Dinah' my spaniel, equally embarrassed on the other; she was overlooking half a dozen of her new-born puppies, which had been removed two or three times from her, and her anxiety was excessive as she tried to find out if they were all present, or if any were still missing. She kept puzzling and running her eyes over them backwards and forwards, but could not satisfy herself. She evidently had a vague notion of counting but the figure was too large for her brain. Taking the two as they stood, dog and Demara, the comparison reflected no great honor on the man."[20]

Conant presents a hierarchy of numeration systems and capacities to comprehend number, which is worked out in impressive detail. "We" moderns at the top have need of a numeration system only as an aid to memory.[21] The Japanese numeration system "is the most remarkable [he has] ever examined, in the extent and variety of the higher numerals with well-defined descriptive names."[22] Below this, the Russians, Welsh, and Catalans are somewhat ahead of Eskimos, North American Indians, and Pacific Islanders. While the Africans lie only just up from South American Indians at the bottom of the scale, the Australians with their mere two or three number words only just make it into the hierarchy, and into humanity. Such a fixed and detailed framework of interpretation would render many of Conant's assessments amusing a century later were it not that many contemporary analysts share his frame of analysis, presenting their rather similar hierarchies in more sanitized terms.[23]

Conant comments extensively on Yoruba numeration, noting that a

species of numeral form, quite different from any that have already been noticed, is found in the Yoruba scale which in many respects is the most peculiar in existence. . . . the words for 11 and 12 are formed by adding the suffix -la, great,[24] to the words for 1 and 2 etc. . . . The word for forty was adopted because cowrie shells which are used for counting were strung by forties; and igba for 200, because a heap of 200 shells was five strings, a convenient higher unit for reckoning. Proceeding in this curious manner [it was reported by a Yoruba that] the king of the Dahomans, having made war on the Yorubans, and attacked their army, was repulsed and defeated with a loss of "two heads, twenty strings, and twenty cowries" of men or 4820.[25]

For Conant, the Yoruba system is "most peculiar and curious" because it does not fit with his hierarchy. He seems in fact to be quite put out by it.

[T]he development of a numeral system is [not] an infallible index of mental power, or of any real approach to civilization [since] a contin-ued use of the trading and bargaining faculties must and does lead to some familiarity with numbers sufficient to enable savages to perform unexpected feats in reckoning. Among some of the West African tribes this has actually been found to be the case; and among the Yorubas of Abeokuta[26] the extraordinary saying "You may seem very clever but you can't tell nine times nine," shows how surprisingly this faculty has been developed, considering the general condition of savagery in which the tribe lived. There can be no doubt that in general, the growth of the

number sense keeps pace with the growth of the intelligence in other respects.[27]

It is Conant's study in which the colonizing intent of the study of
numerations is most evident. Like many curators, Conant wants his exhibition to be tidy, but he has problems that stem from the diverse origins
of the objects he is displaying. Not only were there glaring inaccuracies
in what was collected, but provenance was not properly attended to in
the act of collection. Many collectors of numeral scales (Conant dismisses
them as "vocabulary hunters") were content to collect mere words,
making no systematic inquiries as to original meanings of the number
names.[28] I take it that Conant's complaint here points to a tension in
the colonial enterprise noted by Thomas: "a tension between a scientific
controlled interest in further knowledge and an unstable 'curiosity' which
is not authorized by any methodological or theoretical discourse, and is
grounded in passion rather than reason."[29] As Thomas goes on to point
out, the "giddy passion" of collecting and the scientific ordering tended
to coincide with a difference in placement in the colonial enterprise. The
sailors on ships in the Pacific[30] and the "vocabulary hunters" of Africa
were perhaps strongly moved by an ambivalent curiosity. A desire to contribute to the scientific imperialist ordering of those at home collating
the objects was only part of this. The need to assess the degree of development of the object of their desire, with respect to its capacity to represent
a universal category, did not mean much to many of those sufficiently
moved by the passion of curiosity to withstand the discomforts of collecting. Those impressed by the need for methodological classification were
not much inclined to move from their comfortable studies. We see again
the dual moments of the colonial enterprise.

The catalogue of number systems, and by implication races, assembled
by the likes of Conant, Mann, and Toy can perhaps be seen as the zenith,
or nadir, of the nineteenth-century European sense that a causal science
was about to produce the final answer to questions of "the nature of man."
By the century's end, this collective hope had given way to a sense of
crisis over reason in European thought that was to deepen as the twentieth century wore on.[31] From this crisis, both universalism (in the form
of positivism) and relativism (in the form of social constructivism) reemerged with renewed vigor, only to be outdone by postmodernism. The
first two of these intellectual movements continue the foundationist habit
of hiding the intellectual work of translating and objectifying. The third,
in recognizing it, retreats to the comforts of discourse.[32]

Studies of Yoruba Number by Modern Yoruba Intellectuals

Contemporaneous with these emergent European intellectual move-ments were projects in which Yoruba intellectuals used accounts of Yor-uba numbers. We can understand these endeavors as at once promoting and opposing colonizing. Writing in the city of Oyo, in what would later became Nigeria, at almost the same time that Levi Conant was writing in Worcester, Massachusetts, the Reverend Samuel Johnson, a Yoruba cultural broker of some significance,[33] presents Yoruba numeration as an important element in the definition of "Yorubaness." Johnson was born in Sierra Leone of Yoruba parents, arriving at the age of eleven with his father, a Saro (a returning Yoruba slave), in Abeokuta in what is now southwestern Nigeria. He is recognized as the first modern chronicler of "the Yoruba." His *History of the Yorubas from the Earliest Times to the Beginning of the British Protectorate* was part of the cultural project gener-ating "the Yoruba." This project can be understood as linking the emer-gent British colonial enterprise in what was to become Nigeria with the associated, but separate, formation of "the Yoruba," growing only partly under its influence.[34]

The Yoruba are defined by Johnson first by their place: land held be-tween the River Niger south of its confluence with the Quorra in the east, Dahomey (now Benin) in the west, and the Bight of Benin to the south. It is a tableland that resembles "half of a pie dish turned upside down."[35] Second, Yorubaness lies in language, "the genius of which has eluded those who would reduce it to grammar along the lines of an English or Latin grammar."[36] Johnson takes the Yoruba number system as a particu-larly elegant way of evoking natural number. Separating his discussion of number and numerals in his grammar, number for him is to do with the ways singularity and plurality as such are indicated in language use, and in this, Yoruba is unique in his eyes.

In Yoruba neither singularity nor plurality are routinely indicated through grammatical form; however, both can easily be indicated through the inclusion of appropriate terms. Under the subheading "Number," Johnson reports that Yoruba nouns are not singular in form and hence do not take plural forms. There is no article in Yoruba, so when defi-niteness about the notion of one thing is required, the numeral ọ̀kan is included, or "from the context [it] can . . . be known whether we are speaking of one or more than one: when specification is desired the demonstrative pronoun àwọn (they) or wọn (them) is used with the words."[37]

The Yoruba numerals, in generating a recursion, are strikingly unique, he suggests. While Conant has Yoruba numeration as a "kludge," an ungainly and primitive monstrosity, Johnson views the scale in a positive light, as capturing an essence of Yorubaness. He takes delight in the numerals being "formed on a definite plan, yet more or less complicated."[38] What renders the complicated system of Yoruba numerals identifiable as part of a "definite plan" is an understanding of the grammar embedded in the numerals. Possessing such an elegant technology clearly marks "the Yoruba" as a group of some significance.

In Johnson's hands, Yoruba numbers show the Yoruba to themselves. Some fifty years later, however, Johnson was outdone by S. A. Ekundayo, a Yoruba linguist in the Department of Linguistics at the University of Ife (now Obafemi Awolowo University).[39] Like his Yoruba predecessor, Ekundayo sees Yoruba numeration as a system of elegance and limpid clarity, albeit many others (non-Yoruba) have seen only overdone ornamentation and murky confusion. Having several years earlier completed his Ph.D. in linguistics at the University of Edinburgh,[40] Ekundayo has the Yoruba numeration system as a particular that focuses on, not some shadowy domain of natural number, but a more highly prized universal—the structure of the human mind. His work includes an elaborated theory of the way the working of the human mind mobilizes foundational categories of the real world as the symbolic structures of language. This structure, he argues, is echoed perfectly in the form of Yoruba numerals.

Ekundayo tells us that his is a revolutionary paper, because never before have "Yoruba numeral derivational and representational processes [been discussed] in the context of modern transformational and generative linguistics theory, in particular its [i.e., the Yoruba numeral system's] relatedness to . . . linguistic competence and . . . the relationship between language and mind."[41] Making a strict analogy between the structure of Yoruba nominalized verb phrases (Yoruba noun phrases) and Yoruba numerals, he plots the derivational rules for the numerals. Throwing his lot in with Chomsky, Ekundayo sees himself presenting an example of the rules by which transformations of the "deep structure" of the human mind are generated into language (T-rules, in the jargon of the theory).

For Ekundayo, the Yoruba numeral scale has the appeal of exemplifying a great and universal truth. For him, the particular "the Yoruba numeration system" has no shortcomings whatsoever. It is a perfect example of the class of numeration systems in being a flawless exemplification of the transformation rules implied by Chomsky's theory. Ekundayo spends most of his paper showing just how Yoruba numerals may be thought of as T-rules. With this translation, he is standardizing and objectifying the Yoruba numeration system as a particular—a "set of transformation rules." This particular points back to the universal "the structured human mind."

Ekundayo has set himself a huge task in his paper. His universal "the human mind which lies as a template behind the T-rules" is arcane, being specific to "modern transformational and generative linguistics theory." It is quite unlike the hoary old familiar, "natural number," which almost everyone takes for granted as a universal.

Unlike Johnson, who can rely on his readers' understanding the universal that Yoruba numeration is representing—natural number—Ekundayo must tell his readers both about the universal and about his particular particular. He must evoke for us the structured human mind that lies as a template behind the T-rules, and quite separate from this, residing in quite a different domain, his readers must see Yoruba numerals as a particular a representation of T-rules. He must accomplish complete conflation between two objects, "Yoruba numeration" and "T-rules," and at the same time achieve a complete separation of the particular from the universal.

Those working with orthodox notions of the universal can leave their universal category as implied, present in its absence. We see this strategy in the texts of Mann and Toy, as much as in Conant and Johnson. Ekundayo, having an arcane universal attested by Yoruba numeration, must accomplish the difficult task of showing, and keeping separate, both the general category and the particular he has "found" that attests that category.

Hiding the Contrivance of "Finding"

What do all these foundationist studies of Yoruba number have in common? My text, along with Ekundayo's and all the others, continues the work of colonizing through number. We are "particularising and dividing . . . into manageable parts," in collecting and displaying. The two elements that Appadurai identified in colonial numbering[42] are exemplified in all the texts.

Presenting a relativist analysis of English and Yoruba numerations, I was keen to distance my presentation from the studies of my universalist colleagues. Yet, in sharing the foundationist frame, there is one crucial aspect that my study has in common with theirs. We all assume that we "found" the numerations, the objects we display. As found objects, the numerations project back across the gulf between foundation and instantiation, focusing on the foundation. Were the objects, the systems of number names, to be identifiable as contrived objects, born in a project of collecting and displaying, they would fail to do the work of illuminating the foundation. The elaborate contrivance of "finding" is something we *must* all abide by. It effects a very important hiding: that duplicitous moment where numerals attest numbers *and* numbers attest numerals.

In most of the universalist texts, the hiding of literalization is doubled. The universal to which their found particulars refer—natural number—occupies the unmarked position. This object is so familiar to readers that authors do not even need to mention it as the given entity that their found numerations attest. Of the texts I examined in this chapter, only Ekundayo's and mine must evoke both general category and instantiation. He and I are doing something rather dangerous; our contrivances are liable to show, no matter how tightly we compose our texts.

As scholars, we all fail to notice the work, the translations that achieve standardization and objectification, and through which the Yoruba numeration system, and all other members of the class "numeration systems," are rendered as particulars. These particulars—numeration systems born in a collecting enterprise—are taken to be products of specific societies. We all contend that they exhibit various qualities in variable degrees, but most importantly, we are all convinced that they were found as numeration systems. We assume they were there all the time, ready and waiting for the right eyes. We are likewise convinced that the foundation category, be it "natural number," "mind," or "recursion," was ready and waiting as an ideal, available for us (with the right eyes) to use in our arguments.

Despite the obviousness of the ways numeration systems come into being as particulars, through translations that standardize and objectify, it is crucial for those working within a foundationist imaginary to fail to notice that both the "foundness" and the "objectness" are worked up in our translating. Foundationism can maintain credibility of its visions only by an absolute separation of the domain(s) of particulars and the domain of the universal, or general, category, and it is crucial to abide by the convention of "finding" particular instantiations that attest to the existence of those universal or general categories.

Instantiation and Foundation in My Relativist Study

Along with all my predecessors, I present the particular, the Yoruba number system, as a found entity. My project resembles theirs in that it articulates instances of the class of numeration systems, characterizing them in a descriptive and exhibiting enterprise. Yet, unlike many of them, I have an arcane general category—the notion of a tallying recursion made in a past historical act of tallying. I must show readers the arcane foundation category they are *meant* to see (different logics generated in different sorts of past tallying acts) behind the "found" objects I have assembled for them. I must do this in addition to showing them the properties of those "stumbled across" particulars: the numerations.

In a relativist argument, the general category, the foundation—in this case recursion—and the particulars that are "found" or "discovered" and that attest to this general category are products of the social. At the same time, they are ineluctably distinct. My presentation must on one hand evoke recursion as generically within past social practice of tallying. On the other side, it must also evoke numeration as culturally specific, disparate practices, showing them as found particulars—instantiations of the generic recursion. Importantly, it must hide the traffic by which the boundaries are achieved between the general, historical social practices of the foundation, and the contemporary social practices representing that foundation in the here and now. I effect this conjuring trick in the text of chapter 3 through formalizing the figure of recursion.

Turning my attention to the important question of the origin of the differing forms of English and Yoruba numeral generation, I proposed that "[t]he 'origin story' for zero-the-number is suggestive of the processes by which the social practices of repetitious arranging and patterning in tallying come to be formalized recursively as the domain of natural number, which in turn is represented as a recursive system of symbols." Within the figure of formalized recursion, I locate both "the domain of real number" *and* "a system of symbols." The notion of formalization is strong enough to contain the dual and contradictory moments of the literalization—the simultaneous founding *and* instantiating. A formalized, recursive icon or figure contains and hides the traffic, and the duplicity.

This containment function of formalization reemerges again two pages later: within a formalism, difference is remade as sameness. A profound difference—"in von Neumann's account (and English language numbering) the set 14 has fourteen members, while for Zermelo (and Yoruba language numbering) 14 has one member only"—is located within a sameness that is made by formalizing: "The two formal models agree on the overall structure of number. Each model is demonstrably a recursive progression." Then I blithely deny that difference is significant in any everyday sense: "While this is a significant difference for mathematicians, for ordinary users of number, with no interest in models of cardinality, the difference is irrelevant. For them, that numbers work is all that matters."

Locating difference within an idealized or formalized sameness, and quite outside the classrooms within which my project began, I define and delimit the politics around number. Rendering difference within sameness through formalizing, the politics around number becomes accessible only in "minds," and "discourse," which trade in and with "the abstract." I do not deny "minds" and "discourse" as constituting an important political arena. Yet, as I suggested in chapter 2 when I told of "seeing through" my iconoclastic critique, articulating this arena only partially reveals the politics around numbering, and as I elaborated there, it is in many ways a compromising politics. My literalizing expunges possibilities for difference in numbers in everyday situations like Yoruba classrooms. In confining difference to the abstract and the formal, I deny classrooms like those in my stories in chapter 1 as sites for choice and deliberation over different numbers, effectively denying their workings as an arena of politics around different numbers.

As rhetorical strategy, literalization offends standards of argument, but this is only part of the reason for refusing those arguments as persuasive. Even a moment's reflection on my personal experience of the practical workings of contexts where numbers feature—scientific laboratories, markets in West Africa, Yoruba classrooms, and so on—reveals that I was mistaken in suggesting that what matters in such situations is that numbers work. In any practical going-on with numbers, what matters is that they can be *made* to work, and that *making them work* is a politics. Yet it is a politics that completely evades conventional foundationist analysis, because we have no way of telling the realness of numbers that recognizes them as participants in those contexts. I come back to this issue in chapter 5.

I turn now to complete this decomposing moment by locating the comparative story I told in chapter 3 within the intellectual movements of its time and place.

Formalized Recursion: A Literalized Figure in Which Difference Is Lost

We have seen recursion in the grammars of languages, we have seen re-cursive geometrical trees which grow upwards forever, and we have seen one way in which recursion enters the theory of solid state physics. Now we are going to see yet another way in which the whole world is built out of recursion. . . . We are going to see that particles are—in a cer-tain sense which can only be defined rigorously in relativistic quantum mechanics—nested inside each other in a way which can be described recursively, perhaps even by some sort of "grammar."[43]

Recursion, relativism, and critique, a heady mix with which Hofstadt-er's *Gödel, Escher, Bach: An Eternal Golden Braid* carried the culture of computer hackers into the mainstream in the early 1980s.[44] Along with it came recursion as a pure, singular (if slightly perplexing) object, duly entering my study of English and Yoruba numbers. A copy of Hofstadter's book accompanied me back to Nigeria after a visit to London's bookshops and was one of the few books to be given space in my boxes when I left. Sherry Turkle, finding herself at the center of what she describes as "a make-over" in anglophone public intellectual life, a reimagining of hu-mans as computers, testifies to the special place Hofstadter's book had in this movement.[45]

Bach (who used [recursion] in his endlessly rising canons) and Escher (who used [recursion] in his endlessly rising staircases) are made into cultural heroes just as Russell and Whitehead are made into cultural ene-mies . . . not just because they wanted to ban paradox, but because they belonged to an intellectual culture that saw in paradox problems rather than power. Hofstadter succeeds in getting his readers to sense them-selves as part of a new culture, a computer culture, strong enough to shrug off the culture of Russell, Whitehead, and traditional philosophy and logic.[46]

The *Shorter Oxford English Dictionary,* well behind in this cultural revo-lution around recursion, suggests "recursion" is now rare or obsolete, although in the seventeenth century it implied "a backward movement, a return." Dictionaries of philosophy[47] are more helpful but still miss the bite "recursion" has in contemporary popular culture by citing its arcane twentieth-century origins in mathematics. There "recursive" describes a form of definition or a form of mathematical proof: proofs and definitions that work by containing a disguised self-reference. *The New Hacker's Dictionary* simply defines it as "*Recursion:* See Recursion."[48]

Self-reference, the delight of hackers, is the bane of those seeking to articulate the nature of logic and its role in constituting the foundations of knowledge. Yet it makes recursion particularly appealing and resilient as a foundation category for a relativist project. Recursion dashed hopes for a universal mathematics that flourished in the late nineteenth and early twentieth century: in 1931, Kurt Gödel found an acceptable way to use recursion to show that there is at least one proposition in arithmetic that cannot be shown as either true or false.[49] Some suggest that this served to deepen a more general "crisis over reason in European thought."[50]

In the hands of computer programmers, recursion became an everyday tool: self-reference is what enables complex programs. Turkle argues that Gödel's erstwhile esoteric propositions, and hence recursion, entered popular consciousness as emblematic of computer culture's "heroic triumph" over traditional logic, thus adding to its attractiveness. Turkle suggests that the reification of recursion in computer culture expresses a moral order that has antisensuality, an eschewing of "the flesh things," at its center.[51] Recursion as the world, in an ultimate sense that is what Hofstadter promises in the quotation above, is just what is needed in a culture promising escape from bodies and the mess of the everyday into a pure symbolic realm where one can stay in control. Recursion is the consummate categorical object—it resists decomposition with its celebration of self-reference.

In my relativist project, recursion becomes an adjective: recursive tallying. This use as an adjective characterizes the spread of recursion from mathematics and later computer programming into the social sciences. Tallying, that often messy, thoroughly practical (re)arranging of confusion, entails an endless sorting and placing thus and thus. Tallying is work fit for the menials, what the lower end of empire is assiduously trained to perform. Coupling it with recursion effects both a dignifying and an evacuation of the effort to deal with the ugly, threatening mess. The constant bodily struggle to order is deleted and hidden. The purity of "a behind the mess," a world for our knowledges with their numbers, to be about is ensured, and my failure to grasp difference ensues.

Recursion as a definitive, singular, categorical object is not up to the job I want it to do. It does the job wonderfully for computer programmers, but it cannot sustain a useful critique of numbers. What happens if I resist recursion's categorical pretensions both in its outlaw incarnation and in its savior/hero version? This is to refuse its power to dominate the situations of its use through a single (albeit paradoxical) definition. It is to call recursion's bluff as trope like any other figure, to insist on its multiplicities. It is to insist on looking for number as real and different in places other than minds and discourse. In my next chapter, I return to the inexpert fingers of Yoruba children as they stretch and wind string, and grasp jugs and graduated cylinders, as their bodies are disciplined in the material routines and (re)arrangements of quantifying. Becoming familiar with quite new ways to tell number's realness, I articulate that arena of politics around numbers that emerged in those Yoruba class-rooms in chapter 1. We see that a complexity surrounding numbers can be managed by mucking around with them. That is unthinkable within a foundationist frame where numbers are definitive and given, set and solid, determined by the structure of the abstracting logic by which they are constructed. One cannot tinker about with abstract entities.

Chapter Five

Toward Telling the Social Lives
of Numbers

*I*n 1921, a district officer representing the British Colonial Office in Ibadan received a letter from a Reverend Alexander Akinyele.[1]

[T]he recent 1921 census was not correct on the whole for two reasons: 1. It is a customary instinct in Yorubaland to be disinclined to tell or count the number of people in a town, District or Household for fear they will die—this instinct finds expression in the proverbial saying "A ki ikà ọmọ f'ọlọmo, A ki ikà ọmọ f'obi" "It is not customary to count children for, or in the presence of their parents." 2. The recent census unfortunately coincided with the tribute collection. This makes our ignorant people strongly suspect, though wrongly, that the intention was to extend tribute-payment to women & children. The last census was correct only where the minds of the Heads of Houses had been prepared by men intelligent & well known to them. So, I would humbly suggest:

Either that your respected self or one acquainted with, or wellknown to our people like yourself again undertake to recount all quarters whose Bales were not previously instructed before the last counting; or (to save time and money) call upon the supervisors who did not proceed upon that line suggested to make re-count, with the advice to instruct the Bales before recounting.

The method we adopted in our quarter was to call the Headman

92

with another man & a woman; to ask them to give the *names* of men, each represented by a stone; after all the names were exhausted, the stones were counted; so for women & children. If some such method were adopted with the preparation spoken of, a recount in 2 or 3 adug-boo or quarters would, I presume, show that an increase of nearly 25% would not be an exaggerated estimate.[2]

What can we make of this letter? Perhaps our first response is to empathize with those past citizens of Ibadan over the desire to avoid further taxation, and to wonder if the connection between census and taxation is not more direct than Akinyele suggests. His prescription that the census be carried out with a compound's headman plus another man and a woman perhaps indicates that a headman does not always know exactly who might be living in his compound. It is suggestive of the life of these compounds: crowded living quarters and mobile lifestyles. The implication that the "ignorant" inhabitants of Ibadan households might be tricked by the use of pebbles is more puzzling. It seems that ordinary Ibadan people do not object to a ritual in which a named person is made analogous to a stone, but they fear one in which a named person is made analogous to an uttered number name. Equally confusing is the proverb quoted. What can we make of the notion that children should not be counted in the presence of their parents?

In beginning to consider these puzzles, I point to something familiar: Akinyele's inclusion of an extra step in a numbering exercise. It is reminiscent of Mrs. Babatunde's lesson, which I described in chapter 1. Her introduction of jugs into measuring lung capacity similarly inserted a small "extra" routine into quantifying. Akinyele explains his introduction of the repetitious action with little stones into the numbering routines of the census as getting around the prejudices of "ignorant people." Leaving Akinyele's prejudices behind, I prefer to take his innovation at face value. I recognize that he is put out, disconcerted that his compatriots do not respond to numbering in a modern way, and realizes that he can do something that links their view of numbering with modern number. I am interested in looking at what he does to make numbers work here. I see that his introduction of an extra step has the effect of rendering what in general (Yoruba) parlance is a single unit—an *ilé,* a household—as a plurality. In the routines with stones, what almost everyone knows and accepts as a single whole thing—a household—undergoes transition and becomes "a many"—its inhabitants—but not as a number. Similarly, Mr. Ojo's disconcerting lesson on length makes a single entity, a child's height, into a bundle of strands, and Mrs. Babatunde's actions with jugs make the singularity of lung capacity into several jugs of water.

If we set things this way, we can also see that those "supervisors" who seem to have been satisfied with a number name "plucked out of the air" by a cavalier headman remind us of Mrs. Taiwo's lesson on length. There, children dancing around with pieces of string and rulers seemed to choose numbers entirely at random as they "measured" length. In both these instances, the saying of number does not mesh or link up with the bodily actions; they remain two entirely separate actions. Like Akinyele, we can be certain that those numbers do not correctly generalize.

This suggests that the "doings" of numbers in Ibadan households as part of the 1921 census and the "doings" I witnessed as part of lessons in schoolrooms in the nearby city of Ile-Ife some sixty years later might be linked. Like Mr. Ojo, Akinyele helps us glimpse something other than the purity and simplicity that we normally associate with numbers. We recognize that the numbers that had life at the hands of Akinyele were numbers that fitted into a Yoruba way of doing things and were also numbers that joined up with the ways English-speaking British officials did the work of the British Empire. In just the same way, the mathematics and science lessons I described in my first chapter make obvious an odd sense of contrivance around number. We can imagine the numbers that emerge in the handiwork of Mr. Ojo as linking up with those that have life in the vibrant markets that abound in Ile-Ife as much as they link up with the numbers that emerge in the university laboratory as students do their practical work.

In chapter 2, I proposed an alternative imaginary for analytic work, one that sees worlds as emergent. This imaginary brings with it a framing for telling realness quite different than a foundationist framing. While foundationist stories explain entities like number as determined either by nature (universalism) or by a social past (relativism), I suggested that stories expressing this alternative framing would explain what numbers are in terms of here-and-now routines of practice, ongoing collective acting. These stories are quite different than foundationist stories; for one thing they allow that multiple sorts of numbers were circulating there in Ibadan, both linked and separately. They tell objects/subjects as outcomes of past collective going-on and recognize their participation in remaking particular times and places as (re)generating worlds. Such a telling of number is the work of this chapter.

I begin by elaborating the context of Akinyele's letter, telling a story of numbers in early-twentieth-century Ibadan. My aim here is to focus on the "complex landscape of numbering" that emerged as colonial governance and the Ibadan civic engaged each other. Next, in considering the routines of doing number in a census like the one that so engaged Akinyele, we see that some of number's remarkable capacities lie in its being a relation of unity/plurality. I follow this by exploring how we might credit the strong sense that many have of number as definitive and deterministic. I consider the workings of number as interpellation. This brings us to the point where we can begin to understand the paradoxical notion that numbers are multiple, while still crediting the sense in which they are singular and definitive. Recognizing possibilities for multiple numbers, we start to see how a politics about difference in number might be understood.

Numbers in Early-Twentieth-Century Ibadan

Akinyele's account of his innovations in counting during the 1921 census in Ibadan is particularly useful because this novel way of enumerating inhabitants is set in a place and time when numbers were being made to work in new ways. Numbers were becoming a significant political element in ordinary Ibadan life in ways that stabilized and unsettled both British officialdom and governance through Ibadan's chiefdoms. We can locate Akinyele's odd doing of number within worlds far wider than classrooms, and follow numbers as they make and remake Ibadan and its citizens, and are themselves remade in the politics of an emergent civic community.

The politics of the emergent cultural and social project that was Ibadan at that time has been explored by Ruth Watson in her book *"Civil Disorder Is the Disease of Ibadan": Chieftaincy and Civic Culture in a Colonial City.* She plots a detailed and symmetrical history of Ibadan, drawing equally on diaries and letters of Yoruba chiefs and other citizens and archives of the British Empire and the letters of its servants, supplementing this with contemporary oral histories. Watson traces Ibadan's history from its foundation as a war camp in 1829 through a "long century" to 1939. She elaborates the processes through which Ibadan was conceived as a civic community, showing how they shifted from battlefields where splintering and volatile alliances between chiefdoms were "negotiated" through war, to become fields contested through discourse. Here battles were fought out through judicial practice, commerce in agricultural goods, and trading in imported and manufactured goods: financial transactions, played out with and through British interests. Most importantly, this civic was constituted in and through politics surrounding successions of chieftaincy titles. In these negotiations, "the maintenance of a body of followers remained critical for expressing civic power."[3]

Watson understands colonial governance as a heterogeneous and unstable exercise, not a coherent imposition of bureaucratic procedures. She gives insight into the intimate interactions of British colonial officials as they ran the British Empire, and equally into the ways Ibadan chiefs made the best they could of being a subject people. Alexander Akinyele's letter was written during that period of colonial governance of Ibadan (approximately 1913–26) that is recorded as "the great blot"[4] by one Yoruba historian of Ibadan, and is known colloquially as a time of *Sì fìlà dòbá lè* (Off with your cap!).

In 1913, British administration of Ibadan changed direction as Resident Ross and District Officer Grier acceded to office. Abandoning strategies painstakingly developed over twenty years since treaty negotiations in the late nineteenth century and the signing of the Ibadan Agreement, Ross and Grier set about instituting their own vision. The changes aimed to empower the *Alaafin* of Oyo, extending the political jurisdiction of Oyo, a neighboring city of quite different history and civic arrangement, over Ibadan and its subject towns. In British eyes, Oyo was ruled by a "natural," that is, royal, chief and was "a genuine native town clean and well ordered, while [in Ibadan] everything [was] dirt and confusion."[5]

The chiefs saw things quite differently. Alluding to the Yoruba wars of the nineteenth century and to the widespread understanding that peace in Yoruba country depended on Ibadan, in a letter to Governor Lugard they warned that "it [was] an innovation fraught with consequences for the Alafin to tamper with Ibadan's possessions."[6]

The subject towns were related to Ibadan through the military empire Ibadan had conquered during the Yoruba wars. The towns and their inhabitants were part of the civic status of Ibadan's chiefs, of both economic and social significance—a chiefdom had many parts. The towns made tribute payments to *ajèlè* (agents) contributing to the private stores of the chiefs.[7] Collectively, "subject people" added to the following claimed by a warrior chief—he asserted himself as their *Baba Kékeré* (an official title implying father). From the earliest years of colonial rule, British authorities began undermining this *ajèlè* system as they sought to increase the power of the *Alaafin* of Oyo.

For District Officer Grier, the most significant element in the project of colonial governance was developing "a systematised form of taxation."[8] As he saw it, to establish the native treasuries, the old form of tribute paying had simply to be converted to payment as taxes. Levies previously paid to chiefs in agricultural goods should now simply be paid as money instead. Chiefs could serve as tax collectors and be compensated by salaries for what they must now hand over to the Ibadan treasury. Grier translated a complex transaction fraught with opportunities for political intrigue, to numbers.

In January 1917, as part of this scheme, Ibadan Council of Chiefs agreed to provide an annual sum of £200 to recognize the *Alaafin,* systematizing a nominal payment of kola nuts, which Ibadan had previously given to the ruler of Oyo. Eight months later, Ross coerced the chiefs to agree to raising this to half of the salary of the Oyo king, £4,500. Further, on 29 September 1918, "soldiers armed with their guns" stood by while Captain Ross announced to the assembled Ibadan crowd that every adult male in the town would henceforth be charged a tax of seven shillings.[9] As for the 1921 census, the ensuing tax assessment and collection exercise was, in the eyes of colonial officers, not a resounding success. During the 1920 assessment, the tax officer was highly critical of the chiefs, particularly the *Bálé,* who had come to be recognized as the "head chief" under colonial rule. As far as the tax officer was concerned, the amassing of £15,000 in twelve working days had been achieved *in spite of* the head chief. The point I am making here is not about whether these enumerating exercises were technically flawed. I am trying to point to the new social lives that number was taking on in the Ibadan of those days.

We recognize from Watson's stories that allegiance between a chief and his followers was a negotiated, case-by-case affair. The strength of an *ilé*—a chief's "house"—was not given, solid, and stable. The chiefdoms of Ibadan were neither permanent nor territorially based; they were an ever-shifting expression of allegiance, alliance, feuds, and intrigue. In this volatile environment, the disinclination to enumerate the strength of towns, districts, and households in a census begins to make sense. Numbering attributes and solidifies, and in a situation where what matters is contingency in acting and keeping political options open, this is not a useful direction. We might say that, from the point of view of Ibadanites at that time (both chiefs and followers), enumerating what the British saw as their colonial subjects was to be avoided because it limited capacity for political action. Evidence of a chief's substantial (or meager) following, thanks to British census taking, was now made as capacity of number.

Before British interventions, numbers, in the form of amounts of money and goods, could be used to dissemble, to hide the actual following a chief had acquired. Previously, the strength of a chiefdom was revealed only on the battlefield, where numbers of followers was only one of the relevant factors. After the census, however, the strength of a chiefdom could be expressed in a new way, as a number. Numbers reveal the situation just as starkly as defeat or victory on a battlefield. The revelation effected in the census pointed to the chiefs with the most followers, and for the British this was a good thing.

While they struggled to evade attributions made by the census, the city's chiefs were expected to extract tax revenue from their followers and in a public act hand it over to the Ibadan treasury.[10] The amounts that chiefs agreed to hand over to the *Bálé* could be managed and manipulated. Here chiefs could work on the various elements within their jurisdiction, and hand over what suited them at the time. From the point of view of Ibadan's chiefs, the taxation system was more workable. Whether and how the attributions and revelations made in numbers there helped or hindered their multiple and various agendas was a contingent matter, and they could use numbers in ways they were familiar with to evade British ordering.

It seems that numbering could work both in the volatile field of chieftaincy politics and in the more predictable politics of British governance, but now extra work had to be done to effect this. Yoruba chiefs and their followers, and British petty officials and their supervisors had to do more than they expected to connect things through numbers. Colonial officials had to get their minions to insert extra routines in the process of counting, and visit headmen and arrange meetings with senior members of a household. Ibadan chiefs had to struggle in new ways with their followers, who with their wavering allegiances could also align themselves in new ways.

If we were to believe my old relativist studies, we might imagine that the chiefs and the British officials had different logics in mind as they struggled to work with each other in early-twentieth-century Ibadan. That picture has them working across a gulf between their logics, which they can neither name nor bridge. I suggest that this was not the case, just as it was not the case in the schoolrooms I came to know in nearby Ile-Ife some sixty years later. After reading Watson's stories, we can intuit that different logics of numbering were at work at that time in Ibadan, and that they were embedded in and in turn embedded the politics of that place and time. We recognize that British officials and Ibadan citizens quickly invented routine actions that linked them up in contingent ways. They strung together new routines of acting and negotiated alliances between numbers just as they negotiated other alliances. It seems that different sorts of numbers were made to work with each other as pebbles were picked up and named as persons, and the strongboxes of the Ibadan treasury were (partially) filled by shillings that had been prized out of the contingent and volatile parts of chiefdoms through new sorts of associations.

Numbers as the Relation Unity/Plurality

Censuses such as that undertaken in Ibadan in 1921 were, as Appadurai has elaborated, a mark of rule by a "modern" state (see chapter 4). Remarkably, they seemed to achieve a connection between all places in the empire—including the compounds of Ibadan—and its center. They transported inhabitants, albeit in a highly abbreviated form, directly to imperial headquarters in revealing the empire to itself. In census making, numbers (inaccurate though they might be) are inserted in tables and charts embedded in various reports. As cadastral numbering, a census collects enumerated bodies as inscriptions, leaving actual bodies behind. In part, this intuitive sense of a "carrying off" achieved by the numbering of a census is the basis of the empathy I have with those Ibadanites of the 1920s over their resistance to being counted.

I suggest that this "carrying off" possibility is achieved by numbers being relations. In order to understand this, let me adumbrate its stages as, with a little improvisation, I follow the steps involved in the 1921 Ibadan census. As Akinyele tells it, first there was the repeated making of a *unity*: a person steps forward, or a name is uttered, and a stone is placed. The routine is repeated until the set of people is exhausted. Persons are translated as pebbles. Second, a *plurality* is made as the stones are taken as a collection, which like all collections in this way of numbering is taken to exhibit the quality of numerosity to a particular degree, a degree that can be represented with a number. Third, this number is rendered a *unity*: the population of a compound entered as an object into its place in a chart. Fourth, a further *plurality* is made as the numbers from many compounds are collected, to enable a fifth step, a further *unity*, the population of Ibadan, and so on. In a series of recursive switches between units and their plurals, a child asleep on his mother's back in a compound in Ibadan on a particular day in 1921 enters the ledgers of the British Empire.

Each of these makings of units and pluralities is achieved in repetition of a series of routinized gestures and utterances. When those various British officials did numbering in doing the census, they did it in routine small bodily actions, little rituals. It was a doing of the relation units/plural, and as I will suggest later, the particular routines they engaged in effect this as a specific sort of unity/plurality relation, a "one/many relation."

My point in elaborating this with a detail that is apparently redundant is to reveal what we might call the holographic effect of numbers[11] as effecting their capacity to seamlessly connect a child sleeping on his mother's back in Ibadan with the ledgers of the British Empire. As routines of gestures and utterances, ritualized repetitions, numbers negotiate a scale shift, from the minutiae of family life to the macrosocial category of British Empire. This capacity depends on a figuring, and a recursive refiguring, which emerges with repeated doings with hands and words. In the precision and specificity of some routine gestures, units are continually reaccomplished from a background of plurality, and vice versa.

The series of routine actions that have a sleeping child in Ibadan as enumerated boy in the census are generating a relation, but not a relation in the sense that the child, as son of Kehinde and Babalọlá, is a kin relation—the embodiment of his parents' reciprocal relationship—but rather as a relation that recursively juxtaposes units and their plurals. This story I have just told, with seemingly redundant detail, of the 1921 census suggests that some of the extraordinary properties of numbers lie, not in their being a true representation of a foundation category, but in their being, materially, the recursive relation—unity/plurality.

Most people are used to thinking of themselves as kin, as "having relations" as we say in English. They are probably less familiar with thinking of themselves *as* relation, as embodying the juxtaposition of their parents, who relate reciprocally as husband-wife. Akinyele probably did not think that his actions with pebbles rendered this boy asleep on his mother's back as a relation. Yet, as a relation of unity/plurality, the embodied number/enumerated boy has the remarkable properties that Marilyn Strathern elaborated for the relation, using kin as her material. The relation, according to Strathern, has two significant characteristics: it contains itself within itself and it requires some criterion to complete it. Strathern glosses these two properties as holography and complexity, respectively. In the case of numbers, the criterion defines the unit that in turn enables plurality and in a recursive turn contains the plurality in a further unity.

> [B]eing an example of the field it occupies, every part containing information about the whole, and information about the whole being en-

folded in each part, [the relation] produce[s] instances of itself. . . . [T]hrough relational practices . . . relations can be demonstrated.

[The relation] has the power to bring dissimilar orders or levels of knowledge together while conserving their difference.

[And] what happens when we bring these two properties (holography and complexity) together, when we consider the facility of The Relation to both slip across scales and keep their distinctiveness? In late twentieth-century parlance, our little construct starts looking like a *self-organising device.*[12]

The relation, and numbers as relations are exemplary in this, has the capacity to elide the macro and the micro, rendering them as levels as it does so, all the while leaving the distinctions between the levels intact. Yet, as Strathern suggests, more than this is accomplished; the figuring implicit in the relation seems to have the capacity to insinuate itself into many further relations.

Number Interpellates

I am telling numbers as real through a focus on the banal material practices through which they are done. My hope is that this will lead to possibilities for recognizing difference in numbering and hence to articulating a politics around that difference. If this telling of number's realness is to be persuasive, it must take account of number's capacities for definitiveness—its seeming singularity.

I am suggesting that numbers are located in the embodied doing of rituals with hands, eyes, and words, but if this is so, how is it that they seem to have the capacity to be definitive even in the absence of any bodily doings? In considering this question, I pick up and recycle the Marxist notion of interpellation.[13]

Interpellating is a form of hailing, a hailing or greeting that "drives."[14] Etymologically the term "interpellating" incorporates the Latin verb *pellare, pellere* (to drive). We can imagine interpellation "driving" to effect outcomes. Interpellation, in the way I use it, is as much involved with the production of "objects" as with "subjects," in that both are seen equally as outcomes or effects of myriad enacted associations. Along with this change, the specific Marxist notion of ideology,[15] understood as interpellating individuals as subjects, is remade as the much more diffuse and general notion of enacting. Althusser's ideology is both much more and much less than enacting as I use it, and the individuals/subjects contrast/ connection that he uses to illustrate interpellation by ideology is both more and less than the numerals/numbers contrast/connection I use here to illustrate interpellation by enacting. Yet the arguments he makes about the ways ideology interpellates individuals as already always subjects translate easily as arguments that I want to make to reentangle numerals as numbers. I re-present Althusser's arguments here—suitably translated to do the work I want them to do.

It is common sense that numerals are numbers and numbers are numerals. It is effected with no sign of contrivance. That this does not cause any problems is an accomplishment, an outcome. Obviously, it cannot fail to be recognized that when I speak and write numerals I am using numbers. Problems occur only when I want to use numerals and insist that I am not using numbers. Then I must go to all sorts of trouble to ensure that readers or listeners focus only on words or shapes on paper, which I must again and again detach from numbers as I did in chapter 3. The sounding, or appearance in writing, of a numeral initiates the rituals of its recognition as a number, guaranteeing that numerals are indeed already always numbers.

As I stand before a rickety market stall with tomatoes arranged into small piles, exuding their tomato smell in the hot sun, "ogún" sounds as a pile of five small tomatoes is pointed to. The pointing-at-tomato-pile-while-at-a-stall-in-a-market interpellates, hails, the numeral *ogún* as numbered-tomato-pile. As such, the little ritual hails me as buying subject, the woman on the other side of the display as vendor subject, and the small pile of tomatoes as a commodity—a unity.

In a corner of this page, a numeral manifests. Its appearance-in-just-that-place-in-a-book interpellates the numeral as numbered-position-in-an-order-of-pages-as-book. At work here is specific performance. The routines are material ritual practice that formally effects *specific* material-symbolic participants, objects or subjects: page-in-a-book, priced-tomato-pile, or purchasing—*óyinbó*, for example.

Althusser uses the example of a handshake to present the notion of rituals of recognition; it is a formalizing of individual as subject, as a certain sort of material-symbolic entity, something that matters in the world. "When we recognise somebody of our (previous) acquaintance ((re)-connaisance) in the street, we show him that we have recognised him (and have recognised that he has recognised us) by saying to him 'Hello, my friend,' and shaking his hand (a material ritual practice of ideological recognition in everyday life—in France, at least; elsewhere, there are other rituals)."[16] In a moment of transcendence, the boundary separating persons dissolves as handshake connects, uniting the pair as one—participating in the subjecthood of this time and place: being Frenchmen. The ritual at once both generates and resolves a tension about the boundary, in the same moment both dissolving and (re-)generating the boundary. The two Frenchmen are figured as different—separate individuals—and as the same—sharing a subjecthood with all that entails. To understand the working of numbers in this way, we need to think about how a ritual performance like shaking hands effects outcomes. We intuit that a ritual series of gestures and utterances effect a body as enumerated, in much the same way a handshake accomplishes an embodied person as subject.

Following Althusser's way of doing things, to show interpellating and how it accomplishes number's singularity, I can pretend either a material or a temporal separation. In elaborating through a pretended material split, I propose a performance, a shared little ritual, as hailing or interpellating material numerals—the uttered words or inscriptions—as the numbered, material tomato pile. The contrast here is between the materiality of the falling sound, "ogún," and that of the pile of red fruit. It is the contrast of the numeral as black marks in a corner of a page of this book, and the number of pages in the book, materially the thickness of the pile of paper. This pretense entails that we distinguish for the moment numerals as material, and the stuff enumerated as separate material. Actually the separate materialities ("ogún"/red fruits or "32"/pile of sheets of paper) of the enumerated body—"priced tomato pile" or "book"—exist only as separate in the telling. In the doing, through doing the routine gestures and utterances of buying tomatoes in a Yoruba market, or using a book with page numbers, there is no separation.

Alternatively, in another attempt to illustrate the achievement of number as definitive, I can pretend that enacting functions as interpellating in such a way as to "transform" numerals to become numbers, and "transform" nonenumerated bodies to become enumerated. This pretends that interpellating is introducing figuring into the body. Enumeration "transforms" all numerals to numbered bodies by the very precise operation of interpellating, and likewise transforms nonenumerated bodies to enumerated, I might say. Here I present things in the form of a sequence, as if interpellation has a before and an after and is a form of temporal succession. Actually, these things happen without succession, the enacting and the interpellation of numerals as enumerated object happen in the same moment and are the same thing.

Suppressing those contrived separations of distinct materialities and of temporal sequence amounts to making numerals clear as already always enumerated participants in enacting. Although this might seem paradoxical, it is the material-symbolic working of numbers to effect them as definitive. Using the notion of interpellation, I am imagining a way numbers come to be as embodied specific participants in collective going-on. I am pointing out that, in use, numerals are already always enumerated objects, and as such constantly initiate the rituals of their recognition, guaranteeing them as numbers just as in Althusser's example a handshake guarantees individual as subject.[17]

We need also to notice that a ritual, a series of gestures and utterances, can vary but still achieve approximately the same outcome. In Japan, a bow generally interpellates persons as subjects. Bows and handshakes, as ritual gestures, effect subjects differently, accomplish different subjects. Similarly, we might expect that the little ritual gestures and utterances that enumerate tomatoes in a Yoruba market will not work to effect numbers in a laboratory where Yoruba teachers learn to teach science. The ritual of Yoruba market enumeration effects enumerated bodies differently, effects different enumerated bodies, than the ritual of laboratory enumeration. If children are interpellated only by the ritual of Yoruba market enumeration, then insisting that they enact the ritual of laboratory enumeration, as Mrs. Taiwo did, will fail to effect enumerated bodies.

Yet there is also the possibility for some slack. Rituals can be connected enough in going on. The singularities that are accomplished in interpellating are robust and plastic. Although a handshake in Japan will have different outcomes than a bow, as a bow will effect differently in France, the Japanese and French have come to know enough about each other for a handshake to sometimes pass in Japan, as a bow might sometimes pass in France. Further, a French man visiting Japan might, after some time, suppress the move to shake hands and bow instead. It is likely that at first the bodily movements constituting the bow will only partially interpellate the visitor as Japanese subject. A stiffness, a less than fluid movement reveals, although it might still "do the job"—effecting a Japanese-Frenchman.

Similarly, in a science lesson in a Yoruba classroom, the series of gestures that have numerals as numbered objects might be close enough to ambiguously do the job effecting both Yoruba and English enumerated objects. It may be that in some lessons neither the children nor the visiting lecturer are disconcerted enough for the interpellating to fail.

Where Are Numbers?

I have used the notion of interpellation to show how numbers become as bodies, and bodies as numbers in little rituals. This seems to suggest that numbers are not objects at all. Yet in proposing numbers as effected in the doings of series of gestures, small bodily rituals, I do not mean to imply that we should give up thinking of numbers as objects or things. I do intend, however, that we should give up thinking of numbers as abstract things that exist in *minds*. This raises pressing questions about the forms that numbers, as objects, might take. In dealing with this question, I go back to the stories with which I began my book.

Mr. Ojo's lesson, which I found so disconcerting, was planned as a reperformance of a laboratory exercise the class had carried out under my instruction. For this laboratory class, I had assembled string, meter rulers, and chalk, and I had drawn up charts on paper for each group of students. Calling Mrs. Taiwo to the front, I demonstrated. Then it was her turn. With her hand resting near the top of my head, she let the string dangle over the back of my shoulder until it just touched the floor. Then as I grabbed the end, we stretched it along the floor and with chalk marked out the end and the point on the string she was grasping. Putting the end of a meter ruler at one chalk mark, she made another mark at the end of the ruler and said, "One meter." Lifting the ruler and re-placing the end on that new mark, pointing at each boldly marked interval on the ruler, she counted. "Ten, twenty, thirty, forty, fifty. One meter fifty," she announced. I repeated it more loudly so the class could hear. A ritual is performed. A series of gestures and utterances is carried out just so, and Dr. (Mrs.) Watson emerges as enumerated, and the words, the numeral, emerge as Dr. (Mrs.) Watson. "One meter fifty" is the science lecturer, and the science lecturer is "one meter fifty."

Some weeks later in Mr. Ojo's classroom, a little boy, 'Diran, was standing before the class, becoming enumerated. The string, which a moment before was stretched with the end under his heel and with Mr. Ojo holding it at a point near the top of his head, is now wound around a little card. Mr. Ojo touches each ten-centimeter interval of the string: "Òkan, èjì, èta, èrin, àrùún, èfà, èje, èjo, èsán" (one, two, three, four, five, six, seven, eight, nine). Then holding the bit of string remaining against the gradations on the card: "Mókònléláàádóràán, méjìléláàádóràán, métàléláàádó-ràán, mérìnléláàádóràán [ninety-one, ninety-two, ninety-three, ninety-four centimeters]. Yes, we have ninety-four centimeters. 'Diran's height is ninety-four centimeters." The sequential lifting and announcing of the strings so arrayed interpellates the numeral settled upon, *mẹ̀rìnlélááádọ́ràán* as 'Diran, as much as the boy is *mẹ̀rìnlélááádọ́ràán*. Numerals become as enumerated children, and children present as numbers. The becoming thus is a particular sort of presentation, a particular figuring.

Failure of recognition within the rituals is also possible. Mrs. Taiwo, ordering her children to do *exactly* as we did in the laboratory, found that her lesson failed. Stretching of string, making chalk marks, and extending meter rulers perform a ritual different from Mr. Ojo's, which we might suspect is familiar to Mrs. Taiwo's children. Saying a numeral, like "one hundred and fifty," and pointing as a long ruler is set along a space between two chalk marks hailed the numeral/body as number in the laboratory where all speak English, but does not serve Mrs. Taiwo in her classroom. She found that the ritual failed to deliver on the guarantee that numerals are always already numbers; numbers fail to present as enumerated children, and children fail to become as numbered in her Yoruba-speaking class. Sometimes things go awry, and disconcertment is the effect of the hailing; an out-of-place ritual fails to materialize number.

At the end of the performances, 'Diran and I are numbers, along with also being "class members." There is a significant partiality about our being numbers and about being "class members." The partiality comes in the distributedness of numbers (and class membership). Both Yoruba numbers and English numbers are distributed as what I have been calling figured materiality. Asking where numbers are helps us see them as bodies, but not just one body and not wholly a body.

As embodied subject, I am figured in many, many ways. The number under the heading "Height" in the front of my passport along with my image and my name partially constitute me as subject, as I partially constitute number. As objects, numbers need other objects to complete them, bodies, for example. Human bodies, computer bodies, and the corporate body of mathematicians, to name just a few. Of course, in turn, being figured as body is also an outcome of other rituals quite separate from rituals of numbering, in the case of the human body, rituals that are carried out both in private and in public.

Numbers are uncompleted, partial, and distributed, located in their performance. We can think of them as everted in performance. We could think of them as being there as complete objects only momentarily, ephemerally, in an accomplished performance. As enduring objects, they are located through matter and across space-time—to use a conventional way of talking about our times and places. Similarly, the realities that numbers objectify are multiple, incomplete, infinitely partial, distributed, and potential. Reality is no longer completed, singular, and given. As I said in chapter 2, we need patience to learn to deal with objects in this new imaginary.

Exploring the Possibility of Multiple Numbers

Recognizing that the ritual of Yoruba market enumeration differs from the ritual of laboratory enumeration amounts to admitting different sorts of numbers. This means that, for example, multiple numbers were circulating in the collective acting that was emergent as civic Ibadan in the early twentieth century. When I write that numbers are multiple, I am *not* referring to number's capacity to order many oranges as one kilogram, or many kola nuts as one basket. That is the contrast of unity/plurality I explored earlier. What I am suggesting is that there are multiple ways to do the relation unity/plurality; hence there are multiple sorts of numbers.

In exploring the notion that numbers are multiple, I use the work of Mol, who shows a multiplicity of objects as constituting a disease—for example, anemia or atherosclerosis.[18] Medical textbooks define anemia: a pathophysiological description of blood hemoglobin levels. This is a clear-cut, categorical statement about a singular entity, one rather like the simple singularity usually associated with number. When Mol goes to doctors' consulting rooms and to laboratories, however, she finds multiple ways of doing anemia and multiple versions of normal hemoglobin; the singularity dissolves. "The reality of anemia takes various forms. These are not perspectives seen by different people—a single person may slide in her work from one performance to another. Neither are they alternative, bygone constructions of which only one has emerged from the past—they emerged at different points in history, but none of them has vanished. So, they are different versions, different performances, different realities, that co-exist in the present."[19]

Similarly, when you look at the work with fingers and implements, machines and images, what we might call embodied work, we recognize many ways of doing atherosclerosis.

> Atherosclerosis isn't one but many. This diversity may be taken as a matter of different *aspects* revealed, or different *meanings attributed*. . . . I argue . . . that atherosclerosis is *performed* in a variety of ways, or, better that the name "atherosclerosis" is used for different objects—which also

have names of their own: claudication, thickening of the intima, loss of lumen, pressure drop, plaque formation, and so on. They differ. The material manipulated, the concerns addressed, the reality performed, all vary from one place to another. The ontology incorporated and enacted in the diagnosis, treatment, and prevention of atherosclerosis is multiple.

Meanwhile a single "atherosclerosis" is often mobilized or cited. . . . this requires a lot of effort. The projection of the virtual object "atherosclerosis" behind the variety of "atheroscleroses" performed depends on the ability to make links between one local atherosclerosis and another. . . . it is often difficult to make such links.[20]

If you go to textbooks, or listen to authoritative statements about the incidence and treatment of atherosclerosis, those same people who do the disease as many with their hands, and talk of it as many when they are doing differing versions, speak of, or figure, atherosclerosis as a single, even pure thing. I suspect that, in a similar way, someone like Alexander Akinyele would delight in the simplicity of number as just one thing, at the same time as he comfortably does this, that, or the other set of actions as number, and even talks in ways that recognize difference between numbers in Ibadan's markets and in the Ibadan treasury. In contrast, I, who have never before managed such different numbers in a professional context, am disconcerted when I find that such management is a significant part of being a science-education lecturer in a place like Ile-Ife, Nigeria.

What Mol is focusing on is something we have always known is there with respect to things like atherosclerosis and anemia. It is entirely a matter of course that they might be done in different ways, so they are multiple, while still being singular. We are entirely familiar with how to manage both the oscillation between the multiple versions done in work with hands and eyes and talk of the embodied work, and between this multiplicity of "body work" forms and the singular entity we deal with when we figure anemia or atherosclerosis as *a* disease. We are quite comfortable with the ways the differing anemias and atheroscleroses relate to each other and to the sometimes convenient understanding that they are a single definitive entity. We routinely think of anemia as a complex object. The complex—the disease—is effected, or holds together, well enough in differing times and places, often in quite various ways. The complex object and the singular definitive object flicker contingently.

Mol extends her insight over multiplicity to contending that the multiple anemias constitute an "ontological politics." Seeing that helps her in asking alternative questions about managing health care, and in recognizing them as political and moral questions. The multiple anemias or atheroscleroses are not all equal. Some are much more expensive than others; some can be done in many places while others can only be done in a well-equipped hospital laboratory. The question of which anemia or atherosclerosis is to be done in what situation at least sometimes seems to be a site of deliberate choice; thus it seems to be a politics. Yet to say *that* means we must rethink the notion of politics as much as we must rethink the notion of ontology.

In bringing Mol's insights in here, what I am suggesting is that the same applies to numbers. For numbers, however, both the multiplicity and the oscillation between number's multiple forms, and the singularity we respond with when we are interpellated by number are much more difficult to recognize. Number is the relation unity/plurality effected in rituals, and it seems that this relation can only be of one type; if you are an English speaker, you are very likely to think that it could only be a one/many relation—the sort of relation we saw working in the Ibadan census. The seamy complexity that is relatively easy to recognize in the diseases anemia and atherosclerosis disappears behind the formalism of the unity/plurality relation, which appears only as a dazzling, single purity.

Multiple Rituals

I suggest that we can intuit the existence of number as simultaneously both multiple and singular in Akinyele's account of what he did in Ibadan to help in the 1921 census. I also recognize it in my accounts of how I managed the singular/multiple numbers (the singular and multiple versions of the relation unity/plurality) in Yoruba classrooms some sixty years later.

I begin by going back to chapter 3 and my study of Yoruba and English numerations. I reconsider the contrast that emerged when I took the two recursive systems of number names back to serial pointing at the set of fingers and toes. In my old text, I understood that process as an ancient process that made a symbolizing. In this performative account of numbers, it takes on a different significance.

The contrast of figures variously contained in English and Yoruba numbers seems to me and to others to be enlightening, yet in my old foundationist frame, understanding its significance in number *use* was problematic. There it emerges as a difference in the form of cardinality, which implies the possibility of alternative ways of symbolizing. I could only recognize it as part of an individual's "psychological makeup," which might perhaps be expressed in some weak notion of how they "see" the world. With a new account of what numbers are, I can now begin to see how the contrast is useful in diagnosing my disconcertments in Yoruba classrooms.

In chapter 3, I noted that

> we might say that the Yoruba language rules for numeral generation can be taken as modeling the following sequence of specified actions with hands and feet, fingers and toes. Beginning with the finger/toe complement of a person gives the major base *ogún* (twenty). Shifting from one vigesimal to the next codes for "starting a new set of fingers and toes"—literally "placing out a new set of twenty" in Yoruba; for example, sixty (20×3), *ọgọta*, is twenty placed three times. Specifying *l'ẹwa ó dín* (reduce by ten) signifies "hands only," and the additional specification *l'árùún ó dín* (reduce by five) signifies "one hand only." At this point we have $(-5 - 10 + (20 \times 3))$, *márùúndínláàádọta*. Continuing in this vein, the next specification is "reduce the reduction by ones" $(-4 - 10 + (20 \times 3))$, *mẹrìndínláàádọta;* and along the fingers and so on to the next hand and along its fingers $(+ 1 - 10 + (20 \times 3))$, *mọkọnlélaàádọta*. Next specify "toes (one foot)" $(-5 + (20 \times 3))$, *márùúndínlọgọta;* then "reduce the reduction by ones, working along the toes" to "add the toes of the other foot by ones" $(+4 + (20 \times 3))$, *mẹrinlélọgọta*. Notice that the first *ogún* (twenty) is not specified before starting. Tallying works toward the first *ogún*, not back from it. Presumably, the numerator starts with his or her own fingers/toes. Yoruba numeral generation seems to imply a picture of numbers nested within each other as hands and feet are nested in persons, and fingers and toes are nested in hands and feet.

It seems that in the series of Yoruba numbers names we can glimpse ritual series. A gesturing at hands and feet, fingers and toes. A literal interpretation of number names has us retrieving a performance that we assume historically preceded the "clotting" of the number names as a series. We might speculate that, perhaps for some time, pointing at hand, feet, fingers, and toes always accompanied performance of the names, their recitation in an episode of counting. The same can be said of the English number names. Summing up my etymological evidence, I noted that the performance we "see" in Yoruba number names

> is very different from the picture we get with the base-ten numerals used in modern quantifying and mathematics, say in English. There ten implies the one-by-one pointing to the ten fingers. Twenty implies two sets of fingers. Twenty-one begins again: thumb of left hand held up; twenty-two: forefinger of left hand, up; twenty-three: tall man of left hand, up; and so on to twenty-six: thumb of right hand, up; and on up to thirty, which is three sets of fingers. What we see in English is a linear passing along the fingers until the end, keeping tally of how many sets, and then doing it again. Each finger/integer is equivalent and related to the one before and the one after it in a linear array.

An old series of gestures is recognizable *in* a series of words—the series of number names. A ritual of other times and places is shifted down into our times and places in the words. The "doings" in the words are not lost, in part because they mimic and are mimicked by another series of gestures. In Mr. Ojo's classroom, a series of gestures with hands and string and cards reenacts the gestures of pointing embedded in the number names. The string stretched from head to heel re-presents the digital complement of a person; little card with its wound-up string re-presents its internal complexity. The process of working from this plurality toward the unit that in English we might call 'Diran's height, through the strands and intervals drawn on the card, re-presents a ritual of tallying. We glimpse a ritual still there in the words, and it matches up with what Mr. Ojo does with his fingers, string, and card.

Analogously, we can see that what we did in the laboratory as we revealed heights echoes the procedures I suggested were elaborated by English number names. The string was kept as stretched, and a unit, a meter, analogous to a finger and equivalent to all other meters, just as all fingers are equivalent in this ritual, is used to set centimeters linearly laid out along the stretch. In a serial gesturing of each of the units, a plurality is arrived at. In apparent mutual imitation, number names work within, and within them works another series of gestures, shifted down in the series of words that are the number names. Importantly, the ritual handed down in Yoruba number names is different from the ritual handed down in English number names.

The possibility of failure of interpellation of numerals as numbers reminds us of the banal origins of the figurings induced as entities become enumerated. There was a time before the series of gestures I demonstrated in the laboratory, congealed as what today, in English, we call measuring height. It too was once an innovation, bringing together for the first time what were, in that long-ago, faraway time and place, random gestures. It is now "clotted" and routinely reperformed in many places, although in her Yoruba classroom, Mrs. Taiwo's performance of the ritual is revealed as just gestures—a random set of bodily movements.

Every time such a congealed series of gestures is reenacted, the actions of other times and places are shifted down into this time and place. Enumerated entities are historical objects.[21] The history of the entity "enumerated Dr. (Mrs.) Watson" has long since become opaque, but just as other entities, like the object "numeration system," emerge to change the "ontic landscape," so too once-enumerated object emerged as a new sort of entity.[22] The rituals shift down the arrangements, the "making do" that effected a specific sort of management of the complexity of collective going-on of other times and places. Once some ad hoc arrangements were made, they were an outcome of contingently going-on, and were perhaps for purposes far removed from the purposes for which the rituals are rerun in our times and places.

Differing Rituals Effect Different Unity/Plurality Relations: Different Numbers

The mutually supporting rituals we can recognize in English numbering differ from those in the composite ritual of Yoruba numbering. This difference leads me to propose that the numbers generated in Mr. Ojo's classroom were different objects than the numbers generated in my laboratory session. The performance of the laboratory rituals of enumeration achieved my students as a relation: unity/plurality. Similarly, Mr. Ojo's pupils were accomplished as the relation unity/plurality. Yet they were different relations. The latter ritual seems to works from a plurality to achieve a unity, which I gloss as generating a whole from parts. The other works from units to accomplish a plurality. I name this as working from a one to effect a many. The students in my laboratory session and the pupils in Mr. Ojo's classroom emerge as different forms of the relation unity/plurality.

Rituals evoke tensions resolved as a moment of transcendence, where a boundary is crossed and regenerated in the crossing. The boundaries made in the traffic as students in my laboratory class and pupils in Mr.

Ojo's class emerge as particular relations are boundaries separating inside and outside enumerated body. I suggest that these differing rituals effecting the unity/plurality relation accomplish enumerated bodies as different relations in moments of transcendence that transgress in opposite directions. In a movement from inside to outside, in the laboratory rituals of enumeration students are accomplished as the relation one/many. In contrast, in a transcendence effecting a shift from outside to inside, a pupil in Mr. Ojo's class is achieved as the relation whole/part in enumeration.

In doing English number, the effect of enumerating is a seeming to proceed from an inside with an outward direction, effecting boundaries and working a one to a many. The plurality of the number that is my height, for example, is a collection of units of height. This is the outcome of the laboratory enumeration ritual. In our laboratory session where I was measured by Mrs. Taiwo, the ritual of enumeration concluded with the announcement, "One meter fifty." The announcement seemed to reveal my inside as a collection of centimeters. Cumulative and externalized was "height of our science lecturer." I giggled a little, nervous perhaps at having my shortness, something I knew as already there in me, revealed in a formalized way. My discomfort was only momentary, soon dissipated in the playful atmosphere of the lesson. In contrast, in Mr. Ojo's class the ritual of enumeration seems to contain a moment of drawing an outside into a body being enumerated. The series of gestures in Yoruba numbering seems to have as its outcome the rendering of a whole as parts, in an inward movement. In Mr. Ojo's classroom, *mẹ̀rìnlélààádọ́ràán* seems to internally partition 'Diran, who slinks back to his desk, obviously uncomfortable.

If we both experienced some little pain, my discomfort and 'Diran's very likely differed. Mrs. Taiwo's revelation of me differs from Mr. Ojo's attribution to 'Diran. The pain arises in different ways because the numbers we become differ. That we both experience a moment of pain arises in number's definitive capabilities.

To sum up: The figuring in the English numbering carried out in our laboratory and Yoruba numbering routinely achieved in places like the markets of Ile-Ife seem to express different versions of the relation unity/plurality. In the former, unity seems to be inside plurality as a one/many relation. In the latter, plurality seems to be inside unity. Here enumeration seems to render the relation of whole/part. Of course, the effect of number's being a relation is the oscillation between unity/plurality, and this is common to all numbers. Thus, in the working of laboratory number, a one becomes a many, to possibly become a further one and so on, and in Yoruba markets, number oscillates between part and whole. The difference lies in the starting point of the oscillation. What was a whole for an inhabitant of early-twentieth-century Ibadan, a compound, was merely a collection of cohabitants for a British official. What was unitary for me in our teaching laboratory, a unit of length, becomes a contingent part in Mr. Ojo's classroom.

Managing Links between Various Numbers

When we look at the banal gestures with hands, eyes, and words that constitute the rituals of doing number, we begin to see a complexity we do not usually associate with number. We see possibilities for singularity and definitiveness, giving a solidity within which to go on to further things. We also see a multiplicity of numbers: some emerge in laboratories often with English words; others have life in markets around cities like Ile-Ife and are usually embedded within Yoruba words. In stories I have told here, we have seen this complexity being managed. Akinyele takes steps to deal with it, and sixty years later so do my students and I. Recognizing management opens up possibilities for considering contrasting ways it might be achieved. Considering ways of managing number's complexity is considering the politics around number.

Ibadanites of the 1920s were chary of being counted by the British. They were inclined to have nothing to do with the British forms of numbering, their distrust growing out of an embodied understanding of how numbers participated in their Ibadan polity. Disagreeing with many of his compatriots, Akinyele, who had a modern vision, was keen for his compatriots to participate. To manage the problems, he contrived an additional ritual that both linked and separated those numbers that were feared by Ibadanites and those numbers so trusted by the British.

In discussion with some of a compound's senior inhabitants, he first rendered the set of people in a compound as a set of stones, and then, perhaps elsewhere (he leaves this detail to the imagination of the district officer), he enumerated the stones. Alert to the clashes that are possible between English number and Yoruba number, Akinyele translated. He began his translating work for the census by talking with the "headman" of a compound, the unit of Yoruba government, internally sectioned though it is: "The traditional Yoruba housing unit is a compound. The Yoruba word for it (*agbo'lé*, a group of houses) is tersely descriptive of its composition. It is a group of compartments, with no clear-cut divisions built in the form of a rectangle enclosing and facing an open courtyard."[23] No doubt, he understood this as political work.

> The Yoruba system of government for a long time has been concerned with the administration of large numbers of people divided into groups and sub-groups.[24] . . . A compound head was responsible for governing the inmates of a compound. The quarter head (chief) was responsible for regulating the inter-compound affairs of all compounds under his care. All the chiefs in council under the leadership of the Oba . . . regulated inter-quarter matters and those of general interest to the town as a unit. . . . The Oba was the principle, the embodiment, the "symbol of authority."[25]

As Akinyele notes in his letter, the British officials seemed not to recognize the significance for their enterprise of the compound as the unit of the Yoruba polity, despite their meticulous empirical observations on "native administration" duly recorded in their reports. From Akinyele's letter, we infer that most supervisors failed to visit and instruct the *Bálé,* and to show them how they might use stones or other things in a little ritual of translation to first remake the compound as individuated person. The result was a compromised census—for his own reasons Akinyele is concerned about this.

The Ibadanites' "customary instinct" to avoid enumeration, to regard it as threatening, thwarted the British. The headmen of compounds and in turn the chiefs failed to cooperate. Given what we now understand about the inward partitioning moment of figuring implicit in the ritual of Yoruba numbering, we can understand this resistance in a new light. We see that the British plan for counting compounds' inhabitants might easily be experienced by Ibadanites as an invasion, an unwarranted bringing of outside dangers to the family's inside. Resistance seems the proper response for responsible headmen.

The British see a census as a revealing of resources. What motivates them is a concern to see how many colonial subjects they have. They want an unhiding, a bringing forth of all from the hidden and inaccessible quarters of Ibadan. What was an obvious "one" for the British failed to interpellate Ibadan's citizens as a one. Ibadanites did not recognize themselves in the unitary colonial subject who lay inside the British census. Compounds for the British are collections, pluralities. If, in making the census, a supervisor did happen to call at a compound, that was merely a way to get at individual persons so as to count them.

Mr. Ojo's serial gestures in the classroom perform what is recognizable as a number ritual to his Yoruba pupils. It enables a going-on: teaching science in a Yoruba classroom. If I base my response on the prescribed curriculum, I call Mr. Ojo's performance a failure. I outlaw the number. I refuse to accept the links he and his pupils have made. I am managing number's complexity in one way, Mr. Ojo in another. We are doing politics around number.

Mr. Ojo's version of number might never have emerged before he brought it to life. A linking hybrid performance, it had enough of Yoruba numbering and enough of English measuring height to pass for both. Perhaps its novelty partly explains the alacrity with which the children and Mr. Ojo's colleagues took it up. For a short time as our class was doing its teaching rounds that year, what could have been, and were once, just unrelated gestures became numbers in Yoruba science lessons. Generation of enumerated child was routinely achieved by the routines of manipulating string, cards, and words: a sequence of gestures and movements that must be repeated in a specific order and with a precision of placement of this and that. Hands, fingers, string, card, words, and so on just so, a little ritual, and in the right place, every time enumerated child emerges. Mrs. Babatunde contrived a similar, but less obvious, linking, and Mr. Ojeniyi linked the numbers through a ritual contained in the number words he and the children were using. Nevertheless, when a science lecturer is visiting, insisting that the classroom should be a laboratory, the banal, novel ritual can fail.

The objects "height of Dr. (Mrs.) Watson revealed in science laboratory" and "height of 'Diran attributed in Yoruba classroom" are different entities, and they seem to clash. Historically, they emerged in what were once quite separate ontic landscapes. As historical objects, they are still here, and now they both emerge in the same time and place. For a while, both presented in Yoruba classrooms, apparently as options. Sometimes, as we have seen, they can be linked in performance. Quite oblivious to the multiplicity of numbers in Yoruba classrooms, proceeding on the conviction that number was singular and prescribing a specific ritual, I often created the conditions that ruled out linking—until my students managed to alert me. Remembering how long it took me to see reminds me that science's numbers and Yoruba numbers are not equal.

Seeing how Yoruba and English numbers might be linked reminds me also that linking numbers is already familiar work in mathematics and science lessons; for example, the numbers made in counting and in measuring must be managed as both singular and multiple. Perhaps the need for management is first experienced in math lessons when children are initiated into the contrast between cardinal (the value of a collection) and ordinal (position in a sequence) numbers. Children often meet this complexity in their second or third year of schooling. Generating cardinal and ordinal numbers as both the same and different—linking them but also separating them as different numbers—involves making variations in the bodily routines, in the gestures and the words. A different little ritual figures as ordinal number: the gesture that implies collection, for example, drawing a line around some toy fishes lying on the floor, must be absent. Rather, each fish in turn must be pointed to in a linear action and each remains separate. You must say the words differently and precisely: third, *not* three; fourth, *not* four. It is far from easy for most children. Yet many primary school teachers wonder at the difficulty children have. They forget that it is difficult to develop exactitude in gesture. The apparent simplicity of there being both cardinal and ordinal numbers hides the fact that number is not only singular and pure. Staying with the convention of numbers as singular and categorical in the face of obvious differences in their doing often makes mathematics education much more painful than it need be. Sometimes introduction to vectors (a measure in which direction is important) makes things clearer for some learners. There you must go through routines of gestures and utterances that figure objects as Cartesian coordinates as well as serially sequenced. The scalarness of ordinary number must in consequence be made apparent.

Politics of Managing Number's Complexity

The routines that effect Yoruba numbering and English numbering might be managed to link or to separate. Adding to the complexity, we recognize that the links/separations might be managed as a one/many or a whole/part relation.

Rev. Akinyele, Mr. Ojo, and others made connections without making explicit what is being connected and how. They chose to arrange things so that there could be a going-on together. Often in the early days of my teaching in Nigeria, I arranged things so that there was no connecting. The only going-on that I permitted was to do it *my* way. Mrs. Taiwo followed my example in this, and she experienced a failure in going-on. This choosing to arrange things for a going-on together, or a going-on only in *this* way, is a politics. It is a politics built either on trust and commitment to community here and now, or a politics of imposition, a commitment to a knowledge community that is not here and now. This latter strategy amounts to an intimate form of colonizing.

By telling stories of Yoruba and English numbers as I have here, I have linked them in a way that is quite different from the links in practice that I have been narrating. Linking through telling stories effects an explicitness and enables another sort of politics. Such explicit storytelling effects a politics centered on the choice about *how* to go-on; the question can be discussed. This goes beyond the question "With whom should I go on?"

I have told two sorts of stories of numbers and difference in numbers. A relativist version of numbers enables a politics of linking in one way, and the performative account of numbers I have just given suggests another. The relativist making explicit is a paradoxical linking that insists on separation. This is a politics that insists that number's singularity/multiplicity must be managed through a one/many relation. Adopting the figuring of one group involved in the negotiations, it insists that numbers must be maintained as "ones" as purities. It sees the relations between these purities as a form of politics, but it is politics that is solved at the level of the State, not the level of the classroom.

It implies a policy of equal opportunity to evoke both sorts of number in isolation from each other. Children *should* be inducted into both the symbolizing domains, which will offer them access to the separate conceptual schemes. In this politics, those who are authorized by the State and by the community make decisions about how to arrange things in classrooms. Teachers must, with the support of their curriculum advisers, ensure that children practice both forms. That strategy is actually practiced in some places as a form of bilingual education, although not to my knowledge in Nigeria. In that way of managing the complexity of multiple number, classrooms are quite literally divided into two mutually exclusive zones. With foundationist accounts of numbers, the "public problem" of mathematics education in Yoruba classrooms is elaborated definitively, and removed judges formulate a policy to solve "the problem."

A making explicit that adopts a performative account of numbers effects a different sort of politics. A performative account of number like I have given in this chapter opens up possibilities for working a politics that deals with number's singularity/multiplicity in a different way. Here we see the necessity for a politics that works together a one/many and a whole/part relational management. This is a politics that cannot be legislated as complicated procedures of curriculum and pedagogy—policies devised by the State. A commitment to managing number's singularity/

multiplicity in this way is a commitment to working with a complex community. This is what Annemarie Mol calls "ontological politics." The consequence of reshaping ontology that preformative accounts of objects brings has Mol asking where we are with politics: "[If] there are options between the various versions of an object: which one to perform? . . . we would need to ask *where* such options might be situated and *what* was at stake when a decision between alternative performances was made. We would also need to ask to what extent are there options between different versions of reality if these are not exclusive, but, if they clash in some places and depend on each other elsewhere."[26]

These are empirical and contingent matters. They need to be worked out with teachers and their curriculum advisers, and discussed with schools' communities. These tensions cannot be resolved once and for all in a legislative policy determination. A storyteller, a theorist, who struggles to make such things explicit, can help here. Making the multiplicities of objects explicit in terms of how they are done helps to pose some of those questions in useful ways. For a theorist, to do this sort of critique is to participate in the remaking of community.

Part Three: Generalizing

Chapter Six: Learning to Apply Numbers to Nature

A relativist critique of universalist notions of generalizing, the chapter presents a comparative study of Yoruba- and English-speaking children learning to generalize. I report an experiment designed to prove that the generalizing in Yoruba quantifying is as logical and as abstract as that in English language quantifying. The work rebuts the universalist claims of cross-cultural psychologists that African children fail to advance beyond primitive forms of generalizing logic.

Chapter Seven: Decomposing Generalizing as "Finding Abstract Objects"

Decomposing the report presented in chapter 6, I show that its revelation of difference between English and Yoruba generalizing logics is secured by literalizing. This has the effect of denying it as a real and manageable difference. The literalization invisibly holds an elaborate framing in place, and the separations that framing enables legislate a particular moral order. I show the experimental report is comedy: it finds Yoruba generalizing as an illogical form within English generalizing.

Chapter Eight: Toward Generalization as Transition

In an alternative explanation of generalizations, I keep them as objects but refuse their status as abstractions. This leaves me with the vexing question of what objects like "volume" are. Developing my new explanation within an imaginary of worlds as emergent, I first ask how we could understand generalizations and generalizers as outcomes of collective acting. I am then in a position to ask about differences in generalizing. Next, I turn my attention to my experiment to ask about my generalizing about generalizing.

Chapter Six

Learning to Apply Numbers to Nature

*D*isconcerting contrasts begin the respected book *The New Mathematics and an Old Culture: A Study of Learning among the Kpelle of Liberia,* published more than thirty years ago. John Gay, a teacher of science and mathematics at a secondary college in Liberia, and Michael Cole, who held a position in psychology at the University of California, tell a story of a bowl of rice passed among well-schooled young Americans who were asked to estimate how many cups it contained. Most estimates were wildly incorrect. In contrast, Kpelle adults living in central Liberia almost all judged the amount accurately, despite never having been to school. Asked to sort eight cards featuring two or five, red or green, squares or triangles, all the young Americans, without hesitation and in a matter of seconds, sorted them three ways. Yet the Kpelle men and women had great trouble sorting them even one way, taking minutes over the task. Only two-thirds of this group could find two ways to sort the cards, and less than one-third could sort them in three ways. This is anecdotal evidence of what cross-cultural educational psychologists account as a "problem in African thought." According to this set of understandings, Africans reared in traditional societies have trouble generalizing because they are unable to recognize qualitative, that is abstract, aspects of the world.[1]

In differently motivated researchers, such observations are liable to give rise to such questions as "Does the mind of the primitive differ from the mind of technological man?"[2] John Gay and Michael Cole were more careful. Wanting to build effective bridges to the new mathematics they were struggling to introduce, they felt they needed to know more about the logic of indigenous generalizing. Hence, their book describes scientific experiments that demonstrate the problem in narrower terms. On this basis, the authors make certain recommendations about mathematics education in African countries. Such careful and well-intentioned work converts anecdotes like that above and my story about Mrs. Taiwo in chapter 1 into the scientific demonstration of objectively existing problems.

Similarly, Barbara Lloyd worked with Yoruba children, following up many of Patricia Greenfield's studies of the mathematical abilities and disabilities of Wolof children.[3] Their identification of "the universality of delays in operatory development" and the prevalence of "automaton operational performance" rather than a perceptual cognizance[4] in these groups of African children established beyond doubt that, while African children showed capacities to manipulate numbers, they performed very poorly in appreciating qualitative differences. Their studies and others led to the objective demonstration of the problem of a primitive form of quantification in African communities relating to inadequate generalization.[5]

In academic discourse, such objectively existing "problems" are solved through prescription: the mathematics and science curriculum should be this or that. As part of this prescription, it is necessary to identify the cause of the "problem." In this case, a researcher might ask, "Is this African cognitive failure a biological shortcoming of Africans?" If so, it follows that Africans should be left to their own devices in some form of apartheid. If, more optimistically, it is a social disability,[6] it follows that schooling in Africa should be organized to encourage children to abandon the primitive ways of their communities, and this in turn might imply the need for international financial aid to African nations. Settling on a cause leads to the possibility of a solution. That this is also a moral prescription is only rarely recognized.[7] Yet this is clear in the blunt way I have just posed the issue.

Taking the failure of African children to work numbers logically as an objectively existing problem and prescribing solutions, the inevitable outcome, good intentions notwithstanding, is the delegitimation of local forms of knowledge, which are "other" to scientifically produced knowledge. Science understood in conventional universalist ways has no basis for recognizing as legitimate Yoruba, Wolof, or Kpelle knowing and accounts of knowing.

In Nigeria in the 1980s, laughing and groaning by turns as I worked with my students and their pupils, I was offended by this dismissal of Yoruba forms of knowing. My anger fueled a project of developing a theory of many logics, and as part of this, with the help of some of my students, I carried out a quite large-scale experiment in social psychology. The study was designed to show that the logic that Yoruba children learned as members of the Yoruba knowledge community and expressed in Yoruba language and number use, although *distinct* from, was *equivalent* to the logic children learned in a scientifically oriented schooling and expressed in English language.

What follows is an edited version of a paper I published in a journal of mathematics education in the late 1980s.[8] The paper has shortcomings. Perhaps the most serious is its failure to adequately locate the study in terms of the theoretical issues at stake. The motivation for the study is not made obvious in the text, so it is difficult to see the significance of its findings. To rectify this, in an "afterword" to the paper I offer some interpretive comments. I summarize the main findings of the study, explain the aim of the experiment more clearly, and briefly consider its methodology.

Introduction

An English-speaking child learns to chant "One, two, three, . . ." and a Yoruba-speaking child learns "*Ení, èjì, èta, . . .*" Gradually the songs of number names become meaningful tools. What are the meanings that these words take on as they cease to be a mere chant?

To gain insight into the conceptual basis of number usage is not easy. Adults have forgotten how they learned to quantify. Those who use numbers every day are not given to analyzing what it is they mean when they use number. People will talk about numerals and what it is to manipulate them, but not what it means to use number in talk about the material world. Quantification is a process of analogy, and people forget this. Generally, they are not able to recognize the intermediate steps in making the analogy between the things that are said to be in the world and the series of numbers. The most helpful group for the curious person who wants to talk of what it means to use numbers is children. During their learning of number usage, children often make mistakes, and these mistakes can be informative.

It has been common practice among those who wish to investigate children's thinking over the last thirty years to establish dialogues with children. Often discussion will focus on physical matter that is manipulated during the course of the discussion to pose a puzzle for the child. Explaining this puzzle gives the child an opportunity to display behavior (verbal and nonverbal) from which mental mechanisms used in solving the puzzle (to the child's satisfaction) can be inferred. To determine the strengths and extents of a child's conceptual constructs, several related puzzles may be presented, and the manner of their presentation is determined to a certain extent by the child's responses to previous puzzles. It follows that no two children will receive the same interview.

Picture a young English-speaking child who watches as water is poured from one of an identical pair of tumblers. She has previously agreed that the tumblers contain the same amount of water. The water is divided equally between two smaller tumblers, and the child asserts, "Now there's more water in that one [the remaining original tumbler] because it's very tall and these two are low." A Yoruba-speaking child who watched a similar demonstration claimed, "Ó pò nínú èyìí torí pé méjì ni wón sùgbón òkan èyìí" (There is more water in this because this is two and this is one). Both these children are making incorrect generalizations. In articulating their generalizations, they both say that water has features, but they mention quite different sorts of features. The English-speaking child mentions a dimension of space, and the Yoruba-speaking child points to the number of units in which the water manifests. However, neither child construes the feature as an enduring characteristic of water. The English-speaking child assumes that, when the vertical space that the water occupies changes, the total amount of water necessarily changes. The Yoruba-speaking child assumes that, when the number of units in which the water manifests alters, the total amount of water will also vary.[9]

The Children

The first statement was made by Lucy, who was five at the time. She and other members of her family spoke only English. They lived in a small detached house in an inner suburb of the Australian city Melbourne. Lucy's mother was a homemaker, and her father was a factory worker. Lucy spent most of her nonschool life in adult company, or watching television. The second statement was made by 'Dupe, a Yoruba speaker who was six years old at the time. 'Dupe lived in a village in Oyo State in Nigeria; she learned English at school but rarely spoke it outside English lessons. Her parents were farmers, and she already spent many of her nonschool hours working on the farm. She lived in an extended family and passed most of her time in a children's society.[10] Older sisters have tended to 'Dupe's growing-up, as she has cared for those who have come after her. Both Lucy and 'Dupe made incorrect comparisons between "what is here now" and "what was here before," in generalizing.

Lucy and 'Dupe were two of two hundred and forty-four children I spoke with about the features of physical matter that they thought they could take as the basis for generalization. Lucy and sixty-two of her Australian schoolmates were interviewed in English, and 'Dupe and fifty-eight others in her school were interviewed in Yoruba.[11] Sixty-two bilingual children attending a university campus school in Nigeria were interviewed in English, and sixty in Yoruba.[12] For interviewing, our table was usually set up under a tree in the school grounds, and each child was invited to discuss five demonstrations where everyday things were portioned and reapportioned with cups, bottles, bowls, and beam balances. I manipulated water, Coca-Cola, and peanuts, or what Yoruba-speaking children would call *omi* (water), Coke, and *èpà* (peanuts), while asking questions designed to elicit responses that would give insight into which features of physical matter children were using as a basis for generalizing, and how they were using them. Interviewing sessions were tape-recorded. Later each child's responses were transcribed, Yoruba responses were translated into English, and for each child the most telling sentences and phrases were transferred to an index card coded for the child's age and linguistic ability and the language of the interview.

The Responses

The children's responses were classified in several ways. The first categorization was created around whether children made correct generalizations. Classifying each child's responses on this basis, I calculated what percentage of children in groups characterized by age, linguistic ability, and language spoken in the interview made correct generalizations, and I graphed the results.

Monolingual Children

The Nigerian village children learned English at school, but their competence in English was low, and English talk was confined almost entirely to English lessons. For the purpose of my study, I considered them monolingual Yoruba speakers. The English-speaking children came from homes

Figure 6.1. The age-related increase in capacity to generalize correctly in monolingual English-speaking and Yoruba-speaking children.

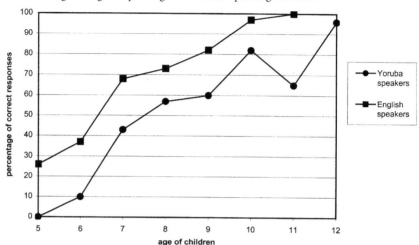

in which English was the only language spoken. They were not learning a second language.

Comparing the percentage of correct generalizations given by monolingual children of differing ages to the age of the children making the response shows that, as they grow, all children learn to generalize (fig. 6.1). Between the ages of five and twelve, the percentage of both English-speaking and Yoruba-speaking children who correctly compare "what is here now" with "what was here before" increases. At around eleven or twelve years of age, nearly all children generalize correctly. The group of Yoruba-speaking children seemed to take longer to always make correct generalizations than did the English-speaking group.

The groups differed too in the type of feature spoken of when the children made correct generalizations. In their responses, the children speaking English talked of two distinct features of physical matter. The responses given by Yoruba-speaking children involved a quite different type of feature. Learning to make correct generalizations was revealed as learning to treat these features as enduring across the little episode of change I introduced during the interview.

Responses of Monolingual Children Speaking English

This is how English-speaking Tony, nine years old, accounted for his generalization that a large flat bowl and a small teacup contained the same amount of peanuts. "That must be the same as that [the peanuts in the bowl and the cup] because that came from there [pointing to one side of a simple beam balance] and that came from there [pointing to the other side], and they were the same before you put them into the cup and into the bowl." Tony is regarding each separated manifestation as an individuated thing. He is allocating the feature of "thingness" to each separate "piece of stuff" before him, and noting that the extent of the thingness does not change. Thingness is a *quality,*[13] a *qualitative* sort of feature. Qualities are imaged as singular linear extension.

Wendy (eleven years) knows too that you can solve the puzzle if you think of any "collection of stuff" as a thing. Nevertheless, she also knows that it is often more useful to construct the feature of volume. Imagine Wendy watching as a small tin, which had formerly contained condensed milk, is filled with peanuts and emptied into a wide shallow bowl, and is filled a second time and emptied into a small glass teacup. I ask Wendy if the peanuts in the bowl and the cup look the same to her and if she thinks they are the same. She replies, "When I look at it one way, they look the same, then when I look at it another way, they don't look the same. That's when I think 'It's just a tin of nuts.' But when I think of the space inside the bowl, I can see that, if I squash that [indicating with her hands the diameter of the surface of the peanuts in the bowl], it will make it higher and I can see that it's the same as in the cup." Wendy is attributing two different features to the collections of peanuts: the feature of thingness ("a tin of nuts") and the feature of volume, and she is noting that the extent of both features remains constant when the cup and the bowl are compared. Wendy could even comment on the usefulness of thingness and volume. I was asking Wendy if she would think that the peanuts in the cup were equal in amount to the peanuts in the bowl if she had not watched when I emptied the tin. She replied that, "if you're thinking about the space that the peanuts fill up, you can try to imagine if they will look the same when the peanuts in the bowl are squashed up the same way as they are in the cup." In other words, if Wendy focuses on the space-fillingness of the peanuts, she can manipulate the images. Volume is another *quality* that matter can have. As qualities, both thingness and volume are usually thought of as intrinsic to matter.[14]

The responses given by English-speaking children who made correct generalizations could be categorized as two types. Younger children mostly pointed to a "thingish" feature, but older children talked of a spatially defined feature, volume. These changes with age are shown in figure 6.2. We see that the percentage of English-speaking children's responses

Figure 6.2. The change, with increasing age, in the type of feature referred to by monolingual English-speaking children who made correct generalizations.

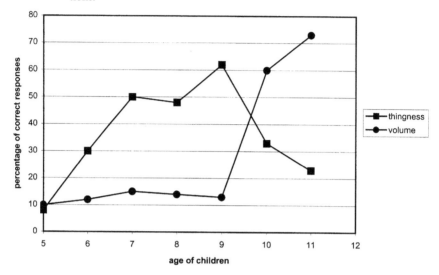

using the feature of thingness increases until children are around nine. After this, it decreases sharply. Talk of volume increases in children older than nine.

Responses of Monolingual Children Speaking Yoruba

When Nigerian village children evoked a feature as enduring, in making correct generalizations, they never mentioned spatiality. Instead they indicated a process involving the definition of a unit and commented on the degree of plurality of the unit. 'Bola (eleven years old) watches as full tins of peanuts are emptied into a plastic bowl and a cup. I ask if there are the same amounts of peanuts in the bowl and the cup (Ǹ jẹ́ iye hóró ẹpà kan náà ló wà nínú kọ́ọ̀bù yìí àti koto yìí?). 'Bola laughs and replies, "Ọ̀kan wà níbí ọkan wà lóhùn-ún" (There is one here and one there). I ask 'Bola if she is quite sure that there is the same amount of peanuts in the two containers; she almost scoffs, "Ẹyọ kan ni ẹyọ kan, àfi tí a bá pin in si méjì bẹ́ẹ̀ ni mo ni wò ọokó pin in" (One is one unless you divide it into two, and I watched and you didn't divide it). Then I ask her whether she would know they were the same if she had turned her back while I poured the peanuts out (Tó bá se pé o wo eèhìn ni gbà ti mo ni da ẹpà náà ni, ò bá mò pé iye kan náà ló wà ninú kọ́ọ̀bù áti koto náà?).

"No," she said, "you might have divided it and taken some away to sell to another person" (Rárá ó seése ki o ti pin in ki o si ti mú díẹ lo tà fún elòmiràn). When I ask 'Bola if the peanuts in the bowl and the cup look like the same amount, she replies that they are not equal: "Wọn kò dógba." When I repeat the question, putting emphasis on "amount" (Ǹ jẹ́ ó dà bi eni pé iye kan náà ni wọ́n), 'Bola asserts that you cannot know whether it is the same amount by looking (Ó nira láti mò bóyá iye kan náà ni wọ́n nipa wíwò).

It seems that for 'Bola a feature that may be called ẹyọ (ìdì), which was translated for me as "bundled unit,"[15] is a contingency of the process of measuring, wọn, by Yoruba speakers. We could imagine Yoruba children making comparative generalizations based on single peanuts rather than collections of peanuts. They would then be said to kà (count) and to be using ẹyọ (ohun). This feature of unicity, ẹyọ (ohun/ìdì), seems to be a feature quite different from the qualities thingness and volume. It is recognizable as contrived. There is no hint of its being intrinsic. It is not a quality of matter but a *mode* of its presentation.[16] The unit is imaged as a multiply divided collection.

The percentage of responses made by monolingual Yoruba-speaking children that indicated that they were utilizing the feature of unicity, ẹyọ (ohun/ìdì), steadily increases as children grow older (see fig. 6.3). By the age of twelve, all the children evoked such a feature in making generalizations.

Figure 6.3. The age-related increase in reference to the feature of unicity by monolingual children speaking Yoruba.

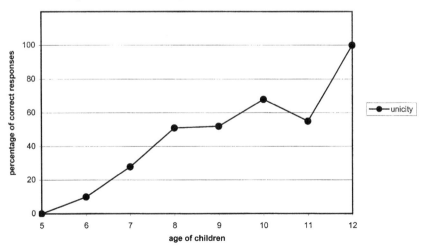

Differences between Monolingual Children

The responses of monolingual Yoruba-speaking and English-speaking children differed in two respects. Considering first the features they invoked, we see that young English-speaking children identify what's before them as a thing regardless of whether it is enclosed by an enduring boundary (as English speakers normally expect a thing to be). They then take the extent of this given quality of thingness (how many things there are) as total amount of matter in making a generalization. Later in their lives, most English-speaking children discard extent of thingness (how many things) as total amount of matter and pick up the notion of three-dimensional spatial extension—volume. They take the extent of the quality of volume as total amount of matter.

Monolingual Yoruba-speaking children do not follow this pattern of development. From the context, they appear to settle upon a unit of manifesting, which may be a happened upon unit *ẹyọ (ohun)* or a bundled up unit *ẹyọ (ìdì)*. The plurality displayed then becomes the total amount.

English-speaking children must learn that the relation between extent of thingness and total amount of matter, and later the relation between extent of volume and total amount of matter, can remain constant, although the look of the matter may change. Meanwhile, Yoruba-speaking children must accept that, once a unit has been allocated, the relation between the plurality of units and total amount of matter remains constant. It seems that Yoruba-speaking children accept the possibility of constancy in the relationship plurality of units and total amount of matter later than English-speaking children accept that extent of quality and total amount of matter can remain constant in the face of changed appearance.

Bilingual Children

The bilingual children were attending a university campus school in Nigeria. All these children came from Yoruba-speaking families, and Yoruba can be considered their first language. However, these children were all English speakers before they began school. They had learned English from their parents as a second language. All lessons at their school were in English.

I classified the responses of the bilingual (Yoruba/English) children with the same criteria that I used for the monolingual children. Figure 6.4 shows the increase in the percentage of correct generalizations with increasing age, for all four groups of children who took part in the study.

Figure 6.4. The age-related increase in making correct generalizations in English-speaking and Yoruba-speaking children

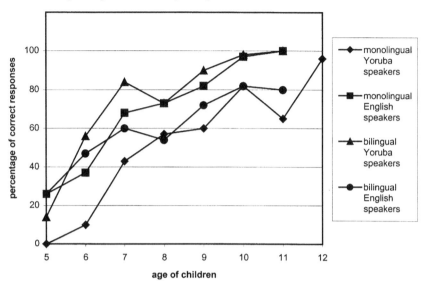

The bilingual children showed a pattern of development toward making correct generalizations like that of the monolingual English-speaking children. This pattern was evident irrespective of whether they were speaking Yoruba or English. The monolingual Yoruba-speaking children lag behind the other three groups of children in this.

Bilingual Children Speaking English

If we turn our attention to the features that the bilingual children mentioned in their responses, we see that, when speaking English, these children spoke of features in a similar way to monolingual English-speaking children and showed a similar pattern of change with increasing age (fig. 6.5). Like the monolingual English speakers, when bilingual Yoruba children speaking English began to generalize correctly, they used the extent of thingness. As they aged, reference to the quality of volume began to replace their use of thingness.

Bilingual Children Speaking Yoruba

In contrast, when speaking Yoruba, the bilingual children clearly spoke of and imputed permanence to a unitary feature, unicity, ẹyọ (ohun/ìdì),

Figure 6.5. The change, with increasing age, in the type of feature referred to by bilingual children speaking English who made correct generalizations

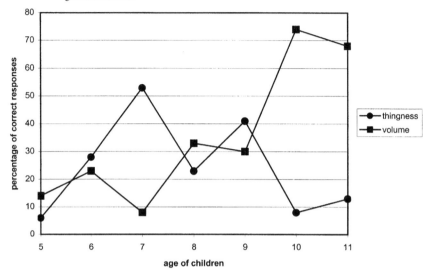

just as monolingual children did (fig. 6.6). The bilingual children did not show the same slow development in this (see fig. 6.4). Moreover, around 15% of bilingual children speaking Yoruba, who made correct generalizations, mentioned spatial situation in their response. Previously, spatiality had been mentioned only by older English speakers evoking the English language concept of volume in justifying their generalizations. It seems that these children were being helpful in describing the nature of the unicity, ẹyọ (ohun/ìdì), they had settled upon.

Here is Folake, age nine, explaining in Yoruba why the Coke in a bottle is the same as that contained in a plastic mug filled with the contents of a second bottle of Coke. "Ara kan náà ní wọ́n tórí pé inú ìgò kékéré náà ni won fi si, o si jẹ́ kí o jọ́ èyìí sùgbọ́n àpapọ̀ èyìí àti èyìí jẹ́ ọ̀kan náà" (They are the same because they put this there in this little bottle and that made them look like this. But the aggregate of this one [indicating the difference in width of the two containers] and this one [indicating the difference in the two heights of the liquid] is the same one). Folake is prepared to comment on the nature of her unitary feature, indicating that it is a unit of space-fillingness, but she still talks of it as a one.

Figure 6.6. The age-related increase in reference to the feature of unicity
by bilingual children speaking Yoruba who made correct
generalizations.

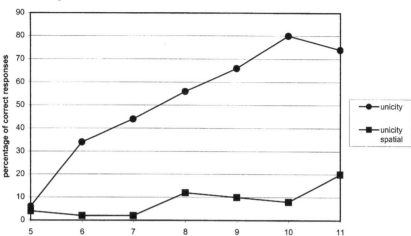

Discussion

Talking of Things in the World in English and Yoruba

How does this account of children talking of their generalizations concerning a total amount help us in understanding how English-speaking
children and Yoruba-speaking children come to use names of numbers
meaningfully in their talk of the world? Frege, a philosopher of mathematics, is helpful here.

> There is the temptation to suggest that we get number by abstraction
> from the things. What we actually get by such means is the concept.
> And in this we discover number. Thus abstraction does genuinely pre-
> cede the formation of a generalization of number.
>
>   The concept has a power of collecting together far superior to the
> unifying power of synthetic apperception. . . . numbers are assigned
> only to concepts under which are brought both the spatial and tempo-
> ral, the non-spatial and the non-temporal.[17]

To understand how children come to use number meaningfully, we
must first understand what "the concept" is in English talk and Yoruba

talk. When we understand the type of abstractions that underlie the children's talk, we will better understand the "natural history" of learning to make these abstractions. We must ask about the origins of "the concept." Is it found in the world, or is it found in our talk of the world?

Like Wittgenstein, I assert that "the concept" is linguistic in origin. Number, according to Wittgenstein, comes not from the world, but from language. We talk of things as composing the world, and we say they have features. Through these features that we say things have, we come to number. "The concept" may be paraphrased as "some feature implicit in the basic unit we talk about as composing the world." It might seem a foolish question, but it is apposite to ask at this point whether English speakers and Yoruba speakers talk of the same type of basic unit when they talk of the world, and come to ascribe features. This apparently foolish question can be answered through a linguistic inquiry. The way to the answer is through considering *singular terms* and their use in talk.[18]

In talk, singular terms name the basic things that are taken as composing the speakers' world. The types of particular things they name are the types of material objects that speakers of a language are committed to saying there are in the world. Identifying singular terms in different languages, and imputing their referents, is a task that is problematic yet possible. Such a linguistic analysis has led me to a conclusion that some may find startling. It is that English speakers and Yoruba speakers usually talk of different types of things when they refer to the physical world. The types of objects that English speakers and Yoruba speakers say they quantify differ.[19]

Using ideas of the inherent characteristics of matter and ideas of space and time, speakers of both languages synthesize ideas of material objects to talk about. But the types of objects they end up postulating are different. It seems that Yoruba speakers use these categorizations in the reverse order that English speakers do. This reversal of the order of applying criteria through which the world is symbolically sliced up results in different types of material objects to talk of.

English speakers talk primarily of spatiotemporal particulars, separated bits of matter set in space/time. In a second-level categorization, spatiotemporal particulars may be taken to exhibit various qualities to varying extents. For example, a bit of matter can be said to have volume to a certain extent. Qualities are abstract objects.

Yoruba speakers talk primarily of sortal particulars, physical matter grouped around sets of characteristics. The sortal particulars that Yoruba speakers talk of can manifest in various modes. In Yoruba language dis-

course, modes name abstract objects, the types of spatiotemporal manifestations that sortal particulars might exhibit. The differences in the types
of material objects, and related abstract objects, talked of are important
when it comes to comparing the ways that English speakers and Yoruba
speakers quantify.

I am now in a position to explain my use of the phrase "features of
physical matter." I choose the word "feature" for its neutrality. An English
speaker who talks of spatiotemporal particulars might say these particulars have qualities. A Yoruba speaker, however, talks of sortal particulars,
and since these particulars have been defined by categorization around
sets of characteristics, these objects cannot be said to have qualities, but
they can be said to have modalities, or modes. The "thingness" and "volume" referred to by English-speaking children are qualities of spatiotemporal particulars. The "unicity"—ẹyọ (ohun/ìdì)—referred to by Yoruba-
speaking children is a mode in which a sortal particular may manifest. A
quality and a mode resemble each other in being features that the material
objects talked of are said to have, that is, in being abstract objects or, in
Frege's language, concepts. In each case, they are concepts through which
number comes to be applied to the physical world. But thingness and
volume are different types of abstract objects than unicity—ẹyọ (ohun/
ìdì).

Identifying a Permanent Feature

The first categorization I made of children's responses centered on the
issue of permanence of a feature of physical matter. Children must learn
to construe the extent or degree of a feature as permanent through some
episodes of perceptual change. This learning is an extrapolation from the
inductive generalization that most infants make concerning the permanence of bodies that they see and feel but do not yet talk of. Children
learn quite early on about bodies. Based on their experience, most babies
seem to make an inductive generalization about the permanence of bodies. Unless they lose sight of a body for some time, they behave as if it
is the same body they are dealing with. Yet behaving as if separated bodies
are permanent is one thing; behaving as if the abstract features things
have are permanent is another.

Features are abstract objects. In construing these abstract objects as
permanent, they must be construed as foreground against a background
in the same way material bodies are construed as foreground against
space-time. In construing abstract objects as permanent, children must

learn to transfer backgrounding/foregrounding from the material to the symbolic.

Different Features

Young English-speaking children who made successful generalizations did so on the basis of assigning the status of "a thing" to the water, Coca-Cola, or peanuts. They allocated an extent of the quality of thingness. For quantifying matter that flows like water, Coca-Cola, or peanuts, a quality that is a continuum like volume or weight is considered more appropriate, but it was only older children who used volume. None of the children used weight, although I invited this by using a beam balance. English-speaking children making correct generalizations talked of a degree of extension and took it as permanent. Irrespective of whether they lived in Nigeria or Australia, children speaking English used the *extent of a given quality* of the stuff before them to connect "what is here now" with "what was here before," to generalize whether the amount had changed.

In contrast, Yoruba-speaking children used the *number of units* of stuff before them to connect "now" with "before." In any quantification context, Yoruba speakers appeared to settle on a unit. This unit may be an aggregated unit *ẹyọ (ìdì)* (for example, a tin of peanuts), or an integrated unit *ẹyọ (ohun)* (for example, a single peanut). On the basis of this unit, a degree of dividedness is allocated as a mode in which the stuff is presented. Yoruba children making correct generalizations talked of a degree of dividedness and took it as permanent.

The Slow Conceptual Development of Monolingual Yoruba Speakers

The Nigerian village children, monolingual Yoruba speakers, appeared to lag behind their bilingual compatriots and Australian children in learning to construe features as permanent. This slow development was not shown by bilingual children speaking Yoruba; they showed accelerated development. An explanation is likely to be found in the circumstances of the children's lives. Village children have a limited experience of quantifying.

Cross-Language Conceptual Transfer by Bilingual Children

In speaking of the unit that forms the basis of quantifying, of the children who made correct generalizations, around 15% of bilingual children speaking Yoruba mentioned spatial dimensions. In speaking of these units as space-filling, it seems that bilingual children are transferring across languages. This transference seems to be associated with a general developmental enhancement. It seems likely that bilingualism enhances children's development in learning the concepts of quantification.

Afterword

The aim of my investigation was to contrast the ways Yoruba-speaking and English-speaking children generalize. I showed evidence that English-speaking children and Yoruba-speaking children develop competence in making correct generalizations within a roughly equivalent time frame. Notwithstanding this similar span of development, I demonstrated that there were significant differences in the ways they explained their generalizations. I suggested that, while English-speaking children explained their generalizations in orthodox qualitative terms, Yoruba-speaking children did not. Against the widespread assumption that the feature evoked by English speakers, extents of various *qualities,* are universals and the only valid basis for true generalization, I showed Yoruba-speaking children confidently making correct generalizations without invoking any qualitative criteria. Instead, I pointed to Yoruba children referring to a *mode* of presentation, which I coded as "unicity."

Correct Generalization in English and Correct Generalization in Yoruba

In the stories of two little girls making mistakes, Lucy and 'Dupe failed to remember in a useful way "what was here before," so that they could correctly compare it to "what is here now." In contrast, Wendy knew the routine and kept the extent of thingness or volume before her. 'Bola resembled Wendy in making a correct generalization, and she said quite explicitly that she knew the way to keep the focus on the number of units.

This learning to get it right, to make correct generalizations, is conventionally taken to be learning the abstract concepts of quantifying. It is assumed that what Wendy has accomplished and Lucy has failed to accomplish is a cognitive shift, from working with crude perceptions of the material world to operating in the symbolic world of abstract entities. At the same time, however, orthodoxy refuses the notion that Bola's correct generalization, when contrasted to 'Dupe's incorrect one, indicates a similar cognitive shift from dealing with "low-level" perceptions embedded in the material to using abstract, symbolic concepts. Orthodoxy denies Yoruba quantifying the status of being abstract because "the one" that 'Bola keeps her sights on is taken to be not symbolic in the sense that "the thingness" and "the volume" that Wendy keeps with her are abstract. This quite generally accepted assessment that Yoruba and other different forms of quantifying are somehow not symbolic, not abstract, and hence not logical[20] was what this paper was designed to challenge. I wanted to ask why and in what ways Wendy's "thingness" and "volume" are abstract, whereas 'Bola's "one" is not.

What denies 'Bola's correct generalization the same legitimacy as Wendy's is the interpretive frame of universalism. This is a commitment to the notion that the structure of the material world (or sometimes the structure of experience of the world) is what underlies the category natural number in a direct way. As we saw in Carnap's version of this in chapter 1,[21] that interpretive frame assumes that the sensed world presents as bounded spatial extensions that endure over time. Spatiotemporal objects have qualities, and making the extent of qualities identified in objects analogous to the extent of the number concept is taken as the sole basis of logical quantifying. *Symbolizing qualities* in making abstractions is the first step toward abstract logical quantifying. 'Bola's way of doing things, presumably representative of the way things are done in the Yoruba language and culture generally, does not engage qualities. In comparing "what is here now" with "what was here before," 'Bola refers to a *mode*, citing a contingent unit of presenting and plurals of this contingent unit.

Redefining and Relocating "the Problem"

Within a universalist interpretive frame, one cannot escape the conclusion that 'Bola and her Yoruba compatriots are working at a low level of abstraction using crude perceptions of the material world. The assumption is that the cultural resources of so-called traditional Yoruba life do not give knowers access to the abstract domain. This enables the inventing of solutions to the objectively existing "problem" that Yoruba knowers fail to recognize the given qualities of the material world. The negative assessments apply to the resources of Yoruba cultural life, not necessarily to contemporary Yoruba individuals, who might be trained to think abstractly. Schooling should encourage children to drop their Yoruba ways and take up rational ways of thought through which one can know and work with the true nature of the real world in a powerfully symbolic way.

Working against such moral and political prescriptions provided the motivation for my study. It entailed two moves. First, the entire process of quantification, beginning with talk, needed to be rendered as symbolizing, and second, the domains of the symbolic had to be shown as plural. These two moves enable my insistence that *mode* is as much an abstract notion as *quality*. It makes possible a challenge to the denial of legitimacy to 'Bola's correct generalization of amount. It provides a basis for arguing that the change we see between 'Dupe's incorrect generalization and 'Bola's correct generalization parallels the change from Lucy's failure to Wendy's success. It says that they are equally moves away from being captivated by impressions of the here and now, to an ability to work in the symbolic domain of the abstract.

The first of these moves, identifying quantification as entirely of the symbolic, was achieved through definition. I summoned support of two eminent philosophers. The second move was achieved with experimental evidence. My paper laid out in detail the type of abstract feature that Yoruba-speaking children evoke in making generalizations about amount. It showed that adoption of this concept by Yoruba-speaking children more or less paralleled the way that English-speaking children took up concepts generally recognized as the abstract concepts of quantifying. In those ways, my investigation suggested a parallel Yoruba knowledge world, which might be regarded as equivalent in worth to that world of "Western" knowledge. I found and presented evidence of these separate but equivalent abstract domains in the children's responses. This is the work of figures 6.2, 6.3, 6.5, and 6.6.

That these differences can be shown I took as vindicating my relativist position. Universalists argue that relativism is radically incoherent since it is paradoxical: the claim that all abstract domains are relative to each other is itself a relative claim. By showing that it is actually possible to demonstrate difference, I felt myself to be adequately dealing with those labyrinthine arguments. Relativism authorizes Yoruba modernity as bilingual, seeing it as a privileged form of modernity. Bilingualism confers cognitive enhancement (bilingual children learn faster—fig. 6.4), and it offers expanded possibilities (bilingual children have access to both worlds—fig. 6.6).

In my redefinition, there is still an "objectively existing problem," but it is a different problem than that defined by universalism. It is located in a different place and calls for a different solution. When quantifying through modes is accepted as abstract and as equivalent to proceeding through qualities, what was a problem in Yoruba psychology becomes a problem in the sociology of education. It is no longer a problem for individual Yoruba knowers who must choose to stop being Yoruba. Relativist analysis relocates "the problem" as one for the community of Yoruba educators: if speaking Yoruba has the effect of propelling speakers into a particular symbolic universe, one with an alternative cognitive style of quantifying, then this should be recognized in the mathematics curriculum in Yoruba schools. Modern Yoruba education should be bilingual, and as part of that program, alternative forms of quantifying should be separately taught and learned in schools.

Chapter Seven

Decomposing Generalizing as "Finding Abstract Objects"

Listening to a hundred or so children using English as they talked about some water or peanuts that I had just transferred from one container to another, I heard them using terms that differed conceptually from those used by many of the children I engaged in conversation using Yoruba. In chapter 6, I told of how, either directly or speaking through translators (my students 'Funmi Oyekanmi and 'Nike Akinola), I questioned around 250 children about the basis of their generalization of such situations. I captured the conversations and arranged for their translation and transcription. From each one, I extracted "evidence" to use as I wrote a report of my study of the nature of these two logics of generalizing. By showing Yoruba and scientific generalizing as equally abstract and logical, my study "disproved" the claim that Yoruba children, like other African children, were "conceptually stranded" in a primitive form of quantifying.

The investigation was well intentioned and careful. Its aim was to redefine and relocate an issue that had been previously identified as a problem with "the African mind." I sought to demolish the claim that Yoruba quantification is primitive, to dissolve a claimed deficit in Yoruba psychology of either social or biological origin. I wanted to remake the problem as an issue in sociology of education and curriculum development. My work was motivated by the conviction that Yoruba knowledge traditions had a significant role to play in the development of Yoruba communities through their schools. I was offended by the seemingly racist conclusions of universalist studies that outlawed Yoruba knowledge from the curriculum. I made claims about children learning to "see" abstract objects as they learned to generalize, which seemed to offer a promising basis for development of a culturally relevant curriculum.

The study sought evidence of the existence and character of two domains of symbolic knowledge in contemporary Yoruba life and proposed each domain as offering alternative forms of the abstraction in generalizing. The report adopted the orthodox foundationist framework, having knowledge as located in minds that can "see through" the messy "outside" world to cognize a hierarchy of "inner forms"—abstract objects. Being a relativist study, it took the origin of those forms to be past practices that gave rise to specific forms of abstracting.

In his universalist account of the generalizing logic in number, Carnap terms the abstract object that interested me as "the comparative concept." As he has it, this is the first abstraction inherent in the formalized generalizing of quantifying. "Seeing" the abstract object enables generalization by removing the focus from physical matter to its (abstract) features, which provide the basis for formally comparing objects by the scaling in numbering. In my relativist report, opting for a more descriptive term, I called this abstract object the "features of physical matter."

In this chapter, I decompose the report of my experiment. I begin by showing how, at the beginning of the paper, the figure of matter set in space and time is used to establish the foundation category "features of physical matter." Then, considering the objects presented in my report as found various instantiations of features of physical matter, I show how these objects are actually contrived in a complicated series of banal material arrangements. The final step in this series is an apparently innocuous translation of children's statements through the figure of matter set in space and time. With this translation, various instantiations of the abstract object "features of physical matter" are revealed as found in the children's replies to my questions. We come to identify this seemingly innocuous translation as achieving literalization of "features of physical matter"—a simultaneous making of both an ideal and the forms of its representation. Through literalization, "features of physical matter" is achieved as an abstract object. Achieved unnoticed within the pervasive and formalized figure of matter set in space and time, the literalizing holds my elaborate and painfully achieved ordering in place as a found order. The translation renders the complicated series of material arrangements and orderings as the "methods" of searching for abstract objects. The object through which generalizing proceeds is removed from the materialities of the conversations, abstracted into minds.

This duplicitous and contradictory literalizing moment ensures the existence of a foundation and the domains of its instantiation, as well as minds outside both those domains. It hides the remaking of the modern, tripartite constitution (see chapter 2), and difference in generalizing is trivialized by being located solely in symbolizing linguistic habits. It is lost as a real difference. The report renders my experiment as comedy, albeit not as entertaining as most comedies. No less than any Shakespearean comedy, it stabilizes necessary separations by superficial reconciliation.

Showing a Foundation Category (Features of Physical Matter)

In a relativist telling, the foundation category "features of physical matter" is the outcome of historically located acts of generalizing. These are supposedly acts that the ancestors of contemporary Yoruba and of contemporary English speakers both made but rendered differently, so today we see different instantiations of the category. In my report, the textual move effecting the foundation category "features of physical matter" is a defining move.

I evoke the foundation category in the paper's introduction by telling of incorrect generalizations by two little children—Lucy and 'Dupe. The two generalizations differ in form, but I ignore that, to use the incorrectness to give an account of the way generalization is achieved through serial observations of a named feature of physical matter. In case the reader failed to pick up the crucial elements from the children's replies that I displayed, I paraphrased what the children said, using the figure matter set in space and time. For example, Lucy said, "Now there's more water in that one [the remaining original tumbler] because it's very tall and these two are low." Having her making observations of matter set in space and connecting her observations across time, setting "what is here now" against "what was there before," I have her as concluding that, when the vertical space the water occupies changes, the total amount of water necessarily changes.

Space-fillingness (volume) is pointed to as a feature of physical matter, and thus I exemplify the definition of the foundation category. A little later, using the incorrectness/correctness contrast in a slightly different way, I point to the existence of the foundational category again, this time by showing that as children grow they naturally learn to "see" the relevant abstract objects and make correct generalizations. This is the work of figures 6.1 and 6.4. What I am emphasizing here in pointing to the defining strategies in my text are the ways that the figure matter set in space and time is constitutive of my definition of "features of physical matter."

Seeing the Ordering Work

I messed around with children, words, water, Coke, peanuts, bottles, bowls, beam balances, tape recorders, translators, transcriptions, index cards, tabulations of numbers, and so on, ordering them all into an almost smooth operation. Nothing would have happened without my energy, my organizing, my bringing and carrying, my telling others to do this and to do that, my arranging by putting this here and that there, saying this and that with a zealous and obsessive bossiness. This energy was a source of amazement to others, and sometimes the occasion for good-natured mockery.

The telling of my experiment as searching for already existing entities—the abstract objects of quantifying—hides a vast amount of laborious but creative work. There were hours and hours of listening to children and then to tapes, writing down what was said, noting the other sounds that tell what was going on. Here too was the skilled work of translating from Yoruba to English. Many, many handwritten pages were produced and laboriously copied and recopied by hand as we generated a format on index cards that would seamlessly re-present the children. Their halting, messy, backtracking and roundabout responses were re-presented. Most of the mess was removed. The messiness of the actual situation was systematically removed, leaving an "order of responses": captured, straightened up, and combed out.

I left out the students 'Funmi Oyekanmi and 'Nike Akinola, who translated during the interviews with the Yoruba-speaking children, acting as my assistants. I left out 'Lola Durojaiye bedridden, paralyzed from the waist down in a traffic accident. She translated and transcribed the taped conversations, rendering them first as neat handwritten pages of Yoruba sentences, then as pages of English sentences that sat easily alongside the pages of sentences from the English-speaking children. I thankfully abandoned pages and pages of English sentences as we extracted particular sentences from the general mass of sentences onto index cards. Then words were left behind, along with much else as I underlined some extracts in red, and in triumph translated this as a particular feature of physical matter. Counting the groups of cards, I produced tables of figures. My graphs—immutable mobiles as Bruno Latour calls them[1]—present "found" instances of abstract objects, leaving the children and the words behind. They are marvels of accumulation, cramming 244 children and thousands and thousands of words into small flat spaces, using only numbers, dots, and lines.

With this creative energy, I contrived an ordered/ordering microworld,[2] a field laboratory within which a reproducible phenomenon could be reliably generated. We can understand the phenomenon,[3] what I collected, as an articulated generalization. In table 7.1, I have listed the material orderings though which the articulated generation was generated, recorded, separated, normalized, grouped, and so on. An ordered/ordering microworld has a specific role in a collective. Its purpose is embodied in material arrangements. The role of the microworld the children entered as they sat down at my table set up under the tree was interrogative. Its

Table 7.1. The Material Orderings by and within Which the Object "an Articulated Generalization" Was Evoked

Interrogative conversations were elicited within these orderings.

1. Children: Categorized by age and linguistic ability with the help of the children's teachers.
2. Questions: Categorized by language of delivery with the help of students 'Funmi Oyekanmi and 'Nike Akinola.
3. Materialities: Peanuts, water, and Coca-Cola arranged with bottles, glasses, cups, and beam balances.

The phenomenon generated and collected was "an articulated generalization."

Each instance of collected articulated generalization was ordered in these acts.

1. Labeling with the age of the child making the statement, the child's linguistic capacity, and the language in which the statement was made.
2. Tape-recording 244 children.
3. Transcribing/translating to handwritten pages by 'Lola Durojaiye.
4. Extracting significant phrases onto index cards, color-coded by age of child, linguistic ability of child, and language of conversation.
5. Separating the collection of cards into two subsidiary groups based on incorrect/correct generalization.
6. Separating the group of correct generalizations according to "type of reference made"; i.e., did this child talk of (a) thingness, (b) space filling, (c) contrived unicity?
7. Separating each group created in 6 according to criteria in 1.

material organization was designed to elicit and evoke. I take this notion from Rouse,[4] who, inspired by Foucault, elaborates an analogy among a confessional, an examination hall, and a laboratory. Rouse is at pains to point to the generative and ordering nature of all three contexts. I carried the equipment for my ordered/ordering microworld in the back of my station wagon: a small table and two chairs, a box with equipment to lay out on the table, tape recorder and spare batteries and cassette tapes, notebook and pencil. I took great care in selecting where to set it up in the playground. Sometimes two of my honors students, 'Funmi Oye-kanmi and 'Nike Akinola, were also part of my microworld. So too was 'Lola Durojaiye, lying back in Ile-Ife with another tape recorder and ear-phones, spare batteries by her bedside, listening and transcribing the children's words from tapes into exercise books, translating the Yoruba con-versations into English. The final set of helpers in my microworld were piles of pastel-colored index cards. Onto them, back in my study, I tran-scribed particular phrases and sentences from the exercise books. With the help of the cards, numbers were generated, and tables and graphs of results devised.

Collecting children, water, and so on as words on tape, which could be turned into numbers in tables of results and eventually plotted as posi-tions in graphs, I present some "found" objects, a "Yoruba language gen-eralizing object" and several "English language generalizing objects." Looking more closely, we see that I actually brought these to life in an elaborate series of material orderings. In a similar way, in chapter 3, "nu-merations" were rendered as objects with certain qualities through a series of translations.

Particular Instantiations of "Features of Physical Matter"

Perhaps the most significant work of the report is the task of rendering the contrasting versions of "features of physical matter" as abstract objects found in children's talk. The entity I was dealing with in my study, an articulated generalization, is made an object within my structured interro-gating microworld. I rendered an object from a formalized conversational exchange, so that, for example, 'Dupe's "ó pò nínú èyìí torí pé méjì ni wón sùgbón òkan èyìí" (there is more water in these because this is two and this is one) became a thing as it fell into my hands as an inscribed index card. I collected 244 of these objects. The objects were tagged as I collected them with the conditions in which the response was made: the child's age and linguistic capacity, and the language of the conversation. The 244 entities were understood by analogy with ordinary things like oranges, spatially bounded objects that endure as such over time. Just like oranges, in a foundationist framing my collected objects can have their attributes or qualities identified, and a collection of objects can be separated, labeled, and normalized on the basis of those qualities. The qualities of the material objects I collected—articulated generalizations—are necessarily abstract objects.

Rendering the Children's Statements through the Figure of Matter Set in Space and Time

Foundationist accounts of the generalizing of numbering have it as beginning with "finding things" made from matter, space, and time. Thus, the discussion section of my experimental report has as its first section "Talking of Things in the World in English and Yoruba." The discussion contends that, while English speakers and Yoruba speakers do not talk of the "same type of 'basic unit' when they talk of the world," they do share a commitment to talking of "material objects," and equally that these categories are derived "using ideas of the inherent characteristics of matter and ideas of space and time." Differences in the "basic unit" are related to what is background and what is foreground. For English speakers, I have the background as space and time, and the foreground as matter, with observing minds somehow standing outside both. For Yoruba speakers, I have things another way—space is found in matter, which is connected across time by observing minds.

Here I am pointing to the central figure in modern accounts of generalizing. If comparison between generalizing as abstracting is to be possible, this figure must be assumed as common to both Yoruba and scientific thought. I offer no evidence for this assumption; the figure matter/space/ time is taken as self-evidently the world. What I am struggling to focus on here is the way the figure of matter set in space and time implanted itself in what I heard the children say. Because I assumed the world is matter/space/time, I failed to notice its work as translation.

The consequences of this final unnoticed step are extraordinary. With this translation, the foundation category is reconfirmed, and the two forms of its representation are "found." Yet that is not all: having thus rendered the "known about" and the various "forms of knowing it," the translation also renders knowers and that very special category of knower—the authority—the author-in-the-text of chapter 6. With the introduction of the figure matter/space/time as a translation device at the end of a series of orderings, the foundation and its instantiations are at once both held apart and made mutually constitutive. It amounts to a literalization where the abstract object "features of physical matter" is made simultaneously as a thing and its representation. This literalizing has the effect of pushing my vast ordering endeavor into the background. The many and varied material orderings within which the phenomenon of children's generalizing came to life became merely the tools by which I searched for and found some examples of abstract objects in their talk.

The Matter/Space/Time Contrasts of My Report

As I see it now, introducing the figure of matter/space/time as the central explanatory device, as I did in my report, leads to metaphysical error.[5] The figure matter/space/time is a metaphysical framing secured by a paradox. It has the status of what the world really is, *and* it is used to frame that world.

The foundationist framing of my experiment as matter/space/time is a triple-level abstracting. First, it sets the water in a glass as foreground against an "empty" background—as a given object to be known about. Applied again, it sets the children as knowers similarly framed: knowing subject and known object against "empty" background. Applied a third time, the "knowing of the child knowers" emerges as an object for the experimenter to know about.

Having used matter/space/time to set the scenes thus, I go on to use it also as an explanatory device, failing to notice that the figure cannot be useful in explaining how times and places might differ because it *is* what all times and places are.

In my report, I devote the first section of the discussion to the alternative sorts of things that make up the alternative worlds of Yoruba and English. English "things" are "spatiotemporal particulars with various qualities," which we can understand as foregrounded when matter is set against space-time. Yoruba "things" are "sortal particulars in various modes," which emerge as foreground when space is set in matter across time—this *sort* of matter, arrayed spatially this way or that, now or then. The figure (matter/space/time) has definitive status—it makes worlds, yet it is also the central explanatory device in the paper, framing and separating the Yoruba world from the English.

The figure is at once world making and world framing, in my report. How do I justify this paradox? The explanatory work of the erstwhile framing figure is made possible by shrinking worlds to symbolic representation in language. The only worlds that exist are the worlds that are talked about. Worlds *are* languages. This has the effect of giving the grammar of languages world-making possibilities, all the while ensuring that it is feasible to comparatively juxtapose the worlds using the grammar of one to elaborate the grammar of the other.

My project of showing evidence of dual symbolizing depends on comparatively setting the world-making frames side by side. To achieve this within a foundationist frame, I have no alternative other than a complete withdrawal into language and symbolizing. Importantly, difference has no place in these shrunken and degraded symbolic worlds. Necessarily, difference exists outside worlds. The only possibility for elaborating difference is to construct *that* world as *this*. The effect is that the Yoruba world can only ever be an echo, some alternative form of that original world made in English (and by extension science). We can understand this as "finding" Yoruba generalizing as already and always inside English quantifying. The former's objects are prefigured within the latter.

Making a Moral Order

The contrast in my text between the quoted words of selected actual children and the numbers in the graphs can be used to focus on another sliding shift: from actual children to generalizing knowers. The children were left behind with their names, their curiosity about this strange woman, their shivering in a cold yard in Melbourne or sweating under the trees in the playground in Toro or Ile-Ife. With their teachers' help, I came up with an order of children, at once material and symbolic but with quite a different materiality (words on paper, not wriggling bodies) and forms of symbolism (children as speakers of particular languages with particular ages, not as 'Nike's daughter, or Taiwo's brother, for example). While many teachers helped me in generating an order of children, it was my agenda, not the teachers', the children's, or their parents'. In creating an order "generalizing knowers," I left behind almost everything that most people would regard as significant about the children. In making choices over what to leave behind and what to foreground, the order of children I created is a moral order and the work of creating that order was hidden in my report.

In treating the responses made by children in many conversations, I contrived two orders to play off each other: an order of children and an order of responses. Juxtaposing the orders, I generated objects—"features of physical matter"—and subjects who know and generalize with and through those objects. Yet, because in a foundationist telling that ordering can only be a searching for preexisting abstract objects, the moral work of generating orderings cannot be acknowledged.

The way I presented all this in a report has me as *finding* that order with its objects ready-made, waiting there for the special investigator with the right observation skills. My having found the objects underlies the authority with which I made pronouncements about the true nature of this and that. The paper establishes its author's privileged view, and her claim to be a legislative authority. That author(ity) ends up with variably constituted symbolizing knowledges over here, which attest a disciplined and socialized reality over there. As we have just seen, however, the two domains are set across a boundary created by its breach. In the process of separation, experimenter and theorist are separated. The obsessive, bossy organizer managing a host of allies is out of the picture; now a removed authority is seen. The frazzled and exhausted experimenter, enmeshed and beholden in complex and ambiguous ways throughout the community with which I have worked, and the removed authority, the theorist who unveils the pure abstraction of foundation and its representations, must assiduously ignore each other. The endless messy mediating ordering work must not pollute the delicate separating.[6] The authority, the author-in-the-text, has a privileged view of both the general and the particular. The place from which authority speaks is set within a further frame where knowledge claims are evaluated according to rules of reasoning.

The Comedy: Its Frames and the Puzzles They Generate

Children speaking Yoruba almost never mention what in English we understand as qualitative aspects when they generalize. In my report, this absence of qualitative aspect is shown quantitatively, in English. The fact that Yoruba generalizing is nonqualitative is "proved" with numbers. In English, using numbers must proceed through a *quality*. What quality does the nonqualitative Yoruba generalizing have? The quality of being nonqualitative! Reconciliation is achieved. Yoruba generalizing is remade as an object through its involvement with English language quantifying (that is, generalizing). The object "Yoruba language generalizing" is qualified as having zero extent of the quality of being qualitative.

The effect is comic. It might also be amusing if reconciliation were conceivable here. If there is a laugh, it is the nervous laugh of someone accepting the bad faith of a confidence trick. The "ha-ha" here is a laugh that reconciles what cannot be reconciled.[7]

In the report, one form of generalization translates and re-presents an "other" form, re-placing the "other" within itself. The turns within turns are as dizzying and comic as a Shakespearean comedy. We can adumbrate the plot of a Shakespearean comedy with its contrived reconciliations within reconciliations,[8] in just the same way that I have laid out the "plot" of the paper reporting my experiment, with its reconciliations within reconciliations. I have "found" Yoruba generalizing within the space of English generalizing.[9]

We might understand the privileged view that has accomplished this comic effect as achieved in retreat to a kind of third frame. This is a position somehow outside the time and place where the children and the interviewer dwell. It is a position that offers an overview of knowers and knowing, and beyond them what those knowers know about. In that initial framing, water sits in glasses, Coke in bottles, and peanuts in tins or bowls. Different sorts of matter, "packaged" in different ways. There is a puzzle in this frame that interests those who inhabit the third frame: Is it the difference between the distinct sorts of things, or is it a difference between the distinct forms of "packaging," that gives rise to the possibility of knowing? This is the disagreement between universalists, who opt for differences located in matter, and relativists, who say it's all in the "packaging."

In the second framing, the water in its glasses, the Coke in its bottles, and the peanuts in their tins and bowls are on a table, and around the table sit knowers, children and interviewers having conversations about what they know and how they know it. Located outside the "stuff" they know about and, as bodies, problematically also "stuff," these knowers make observations and use them in proper ways to know. Collecting conversations about knowing made in this frame, removing them to a third frame to look at them in the light of the puzzle in the first frame, the experimenter, whose coincidental habitation of the body of the interviewer is irrelevant, can claim certain insights, assert some knowing about knowing.

The experimenter who opts for the possibility of knowing arising in divisions within "stuff" known about (the universalist) asserts that there is just one way of knowing. If the conversations she collects show that *this* group of knowers does it differently than *that* group of knowers, then one group is wrong. The one that is wrong will, by definition, *not* be that group to which the experimenter belongs. From the privileged position of the third frame, the experimenter can prescribe a solution that abolishes different ways of knowing.

The experimenter who sees that the possibility for knowing arises in different ways of "packaging" (the relativist) *expects* to see differences in ways of knowing in conversations she has carried into the third frame. Finding a difference, she still has a problem. It can only be a difference made in repackaging. Real difference is unknowable. Made in some lost, past act of "packaging" embedded in a way of life by the framing given in linguistic symbolizing of that life, difference remains outside and hence unknowable.

This character—the relativist experimenter—is the one we are concerned with here. When we put things this way, we can see that what I have already drawn attention to—the trivializing of difference in relativist studies—must be so. Difference cannot be realized. Yet in the trivializing, we glimpse what seems to be an absolute difference left quite untouched, as everything is set to go on as before.

Recognizing the Figuring in Generalizing

Grounded in the actualities of particular moments in specific times and places, generalizing purports to give a persuasive account of general, shared, or common experience. The specific and momentary is transformed into something more robust, something beyond the idiosyncratic peculiarities of the here and now. Generalizing is necessarily a transformation, a transfiguring, a transition. Generalizing is a recomposing through some generally recognized contrast, a mediating figure. Such figures circulate in all collectives or communities. They tend to be so embedded that they are transparent; it is difficult to recognize that they are there. As we have just seen, in telling the children's generalizing, I refigured with the matter/space/time contrast. I failed to recognize this as a figure, however; I took it as a self-evidently given world.

Through the mediation of this figure, generalizing was removed from hands, eyes, and voices working with glasses of water and so on, and transposed to minds to become an abstract object. In beginning to ask how I might extricate myself from foundationism's pervasive and invisible guidance, I want to be able to recognize the figuring in generalizing without abandoning the sense that generalizing carries of conferring stability. I understand this as keeping the objects that generalization seems to be— the objects with which many children showed themselves to be familiar— but refusing these objects the status of abstractions. My hope is that I will be able to develop a notion of generalizing's logic as generative while recognizing its definitive and stabilizing capacities.

Chapter Eight

Toward Generalization as Transition

"Now there's more water in that one [the remaining original tumbler] because it's very tall and these two are low." Lucy, age five.

"Ó pò nínú èyìí torí pé méjì ni wón sùgbón òkan èyìí" (There is more water in these because this is two and this is one). 'Dupe, age six.

"That must be the same as that [the peanuts in the bowl and the cup] because that came from there [pointing to one side of a simple beam balance] and that came from there [pointing to the other side], and they were the same before you put them into the cup and into the bowl." Tony, age nine.

"When I look at it one way, they look the same, then when I look at it another way, they don't look the same. That's when I think 'It's just a tin of nuts.' But when I think of the space inside the bowl, I can see that, if I squash that [indicating with her hands the diameter of the surface of the peanuts in the bowl], it will make it higher and I can see that it's the same as in the cup." Wendy, age eleven.

"Òkan wà níbí òkan wà lóhùn-ún" (There is one here and one there). I ask 'Bola if she is quite sure that there is the same amount of peanuts in the two containers; she almost scoffs, "Ẹyọ kan ni ẹyọ kan, àfi tí a bá pin in si méjì béẹ̀ ni mo ni wò ọ̀okò pin in" (One is one unless you divide it into two, and I watched and you didn't divide it). 'Bola asserts that you cannot know whether it is the same amount by looking, "Ó nira láti mò bóyá iye kan náà ni wọ́n nipa wíwò." 'Bola, age eleven.

"Ara kan náà ní wọ́n tórí pé inú ìgò kékéré náà ni won fi si, o si jẹ́ kí o jọ́ èyìí sùgbọ́n àpapọ̀ èyìí àti èyìí jẹ́ ọ̀kan náà" (They are the same—part of the same thing—because they put this there in this little bottle and that made them look like this. But the aggregate of this one [indicating the difference in width of the two containers] and this one [indicating the difference in the two heights of the liquid] is the same one). Folake, age nine.

These children and many others I spoke to as I tipped and poured are confidently talking of some generalizations. We are inclined to smile at Lucy's and 'Dupe's accounts and suggest that they are not yet fully acquainted with generalizations, confident that they will develop an adequate familiarity in time. I labeled what Tony and Wendy talk about as thingness and volume, and what 'Bola and Folake talk about as unicity. I was interested in telling how the unicity 'Bola and Folake talk about differs from what Tony and Wendy allude to.

In coming up with a another explanation of generalizing, I want to stay true to these children and their confident assertions about what I called thingness, volume, and unicity. I want to respect the familiarity they evidently have with these generalizations. While I am determined to keep these generalizations as objects, I am equally determined to abandon their status as *abstract* objects. To do this, I must free myself from the tripartite constitution of foundationism with its physical world, set against its necessarily abstract knowledge, and its removed knowers. In turn, this involves freeing myself from the assumption that the figure that contrasts matter and empty space and time is self-evidently a description of the physical world. This is not to say that I refuse recognition of the contrast of matter and empty space and time as real. Rather, I insist on recognizing its claim to be the world, and the worlds made through it, as accomplishments.

Abandoning the characterization of thingness, volume, and unicity as *abstract* objects that connect serial observations of matter/space/time, I get around foundationism's problem of needing to invisibly justify the ideal and the physical (the foundation and its instantiation), which can only be achieved through literalization. I no longer need to eject differences in generalization from material, lived experience and insert them into symbolizing worlds. I also avoid the abhorrent invisible insinuation of a particular moral order into my stories.

Refusing thingness, volume, and unicity as abstractions leads me to the vexing question of what *are* these objects that the children so confidently talked about and gestured with? Answering that is the work of this chapter. I give an account of generalization within an imaginary where worlds emerge in collective acting. I image this shift from the foundationism of chapter 6 as collapsing the three levels of my foundationist interpretation of the experiment into a single accomplished domain. The triple framing implicit in the foundationist account zooms inward, all privileged views abolished. The effect of such a rearranging is disappearances as well as appearances. This is dizzying work. Things that seem to be essential for generalizing—"knowers" and "known abouts"—disappear. New sorts of entities emerge.

What immediately disappears is the figure of the theorist as the doubly removed observer. The theorist reenters the body of the woman talking to children, so to speak, passing through the second-level frame of the experimenter. The theorist no longer observes difference in generalizing from the "outside," the so-called God's eye position. The figures of the child struggling to know and the matter she knows about morph as well. Children disappear as judges of change in an observed physical world and as having conversations about their observations and judgments. They no longer formulate certain judgments about abstract objects embedded within their talk of matter set in space and time. Instead, their certainty about generalizing, their status as children with a capacity for correct generalization, and the objects that are their generalizations, are taken as outcomes of their ongoing situation. The same must be said of the criteria of a correct generalization. Similarly, the theorist, the teller of stories about those highly structured conversations, is an outcome. Her certainty about differences in generalizing and her status as expert in the field of comparative quantification are effected in the actual doings of the experiment, in the elaborate orderings she engineered.

I propose this new telling of generalizing as capable of grasping differences in generalizing in useful ways. I do not claim this new account should be accepted because it tells the way generalizing "really" is. It is another description generated in an interpretive cosmology alternative to foundationism. The new description is perhaps counterintuitive, but it is metaphysically canny. I hope that my redescription will provide a basis for justifying studies we might understand as empirical forms of ontology and ethics, studying the "births" and "deaths" of objects, and allow generalizing to be understood as moral and political acting, as much as logical.

First, in considering the generalizations about water, Coke, and peanuts that the children made and told me about, I ask how we can understand these generalizations as outcomes of ongoing collective acting. Then, recognizing the children's considerable accomplishments, I ask how they emerge with capacities as generalizers from that same going-on. I am then in a position to ask about differences between generalizing by children speaking English and those speaking Yoruba. Next, turning my attention to the generalizing of my experiment, I ask about my generalizing about generalizing. The novel imaginary I adopt here in accounting generalization helps me recognize an odd self-referentiality in my study of children learning to generalize. There is an interesting parallel between the generalizing that the children were doing during our conversations and the generalizing I was doing in my experiment, which was constituted by those very conversations. This enables me to explore how my accounts of generalizing and the possibility of difference in generalizing can be understood as outcomes of the collective going-on that was my experiment, and how I might justify giving two radically different accounts of one experiment.

Generalizations as Outcomes

The activities with water, Coke, peanuts, cups, bottles, bowls, and glasses during my conversations with children are thoroughly ordinary and everyday sorts of activities. The tipping and pouring, dividing and adding are repeated again and again as ordinary life goes on, as goods are bought and sold, deals are done or not. They are also often repeated in primary school math lessons—at least in Australia. Yet, while such situations are banal and ordinary, they are also specific. The activities that the children and I had conversations about have specific roles in ordinary going-on.

Generalizations emerge from what I am calling "microworlds": specific materially arranged times/places where rituals, repeated routine performances, occur. The rituals in the banal microworld of exchanging peanuts for money, sharing a Coke, and tipping beans into the pans of a beam balance in a math lesson are not usually labeled as such, yet a series of gestures in handling, seeing, and saying must be done just right. In the same way, a ritual must be done just so. In the structuring of the performance, a vast amount of irrelevant complexity is excluded, and momentarily, ongoing collective life becomes extremely simple.

Imagine a peanut vendor in a market. On her head, she carries a tray that holds a small glass-sided box inside which are roasted peanuts, a pile of pages torn from used exercise books, and an empty milk tin. Each sale she makes is the same sort of episode. We might say that the microworld of a sale of a quantity of peanuts by a vendor to a customer proffering a handful of naira notes, is both ordered and ordering. Similarly, the microworld of two village children sharing a rare and precious bottle of Coke, deciding how to ensure equal distribution of the sweet black liquid, is highly ordered and ordering. Both contexts are materially structured, and rituals happen within them. Each purchase of peanuts is roughly the same—there are some elements that cannot be left out. Every fair sharing of a bottle of Coke is almost identical to every other such sharing.

The inner-city Australian children have probably not experienced such situations as these. They are thoroughly used to the rituals of math lessons, however, with beam balances, beans, water, and measuring jugs. They have been habituated to the structured material microworlds of the rituals of quantifying since their nursery school years.

I am pointing out that the sort of generalization I was seeking in children's conversations in my study always emerges from an ordered/ordering microworld, and most of the children I spoke to had extensive experience of being "ordered about" by these microworlds. The Yoruba village children were the group with the least experience of such ordered/ordering worlds. The responses I elicited from them reflect this.

In introducing the notion of interpellation in chapter 5, I pointed to the ways that ritual both generates and resolves tension. I used Althusser's example of a handshake as a ritual that, momentarily transcending difference, unites two men as collective subject, at the same time as it separates them as differently embodied subjects. Althusser's example of a handshake points to the creating and resolution of a tension in which the generalization "collective identity" is effected.

We can recognize that a handshake effects and is effected as ritual in an ordered/ordering microworld. The participants must stand just far enough apart, not too close, not too distant. Right arms must be proffered across the participants' bodies so hands clutch in a particular manner. During a handshake, it is as if the two men are enclosed in a bubble—the confusing complexity of their separate lives is momentarily excluded. Simplification is achieved in a foregrounding. The generalization effected in the doing of the ritual is internally complicated, embedding in unambiguous ways the often highly ambiguous complexity that the microworld structures in ritual. In the case of Althusser's handshake, the complexity that is remade as manageable complication in the ritual is the contradictory separate yet co-constituted and singular identit(y)ies.

I am suggesting that the microworld of the conversations that made up my experiment, indeed, all those microworlds in which quantification occurs, have a similar interpellative effect. In the series of gestures in which I poured and tipped, divided and added, pointed and named, a tension over boundaries and enclosures was generated and resolved. In doing this ritual, the complexity of the ephemeral nature of boundary/enclosure—the ever-shifting flux of existence that is endemic in any collective going-on—was dealt with.

When I talked to Tony, I used a crude beam balance. From a large container of peanuts, I grabbed a handful and placed them in the wide plastic bowl suspended on one side of the balance. Then I grabbed another handful and placed it in the bowl attached to the other end of the balancing stick. I added a few more nuts, and bent down and squinted at the two plastic bowls to see if they were at the same height. I removed a peanut or two from one bowl. Tony agreed that the bowls balanced. Then I tipped the peanuts from one bowl into a tall glass, and those from the other bowl into a squat teacup. Holding the cup and the glass toward Tony, I asked him if they were the same. Aligning my series of actions and the bounded entities generated and manipulated within them, Tony came up with the generalization—he settled on "a thing" in responding correctly to my question. He implied that those two things—peanut collections in bowls—*were* the same and still *are* although they look different now. He effects an ordering in seeing the collections poured from bowls to glass and to cup as "things"—stuff with a boundary. Things are more complicated than ever-changing stuff, but they are less complex; things are stuff within boundary. In the transition from many confusing "peanut presentations" to "a collection of peanuts," complexity has been negotiated as a (somewhat) complicated thing.[1]

Look at Wendy's account. She comes up with a generalization I called volume. As a generalization, volume of peanuts is internally more complicated than "peanut collection as a thing" ("a tin of peanuts"). Things have boundaries, volume has boundary and dimensions, as Wendy tells and shows us with her hands.

We might say that the confusing complexity of the variety of ways peanuts appear is managed by rendering the material arrangements of the microworld, enacted in its rituals, into the generalization. Generalization expresses ritual enacted in ordered/ordering microworld. Both Wendy and Tony find themselves with solidifying and stable units of peanuts, units of the sort that might be subsequently worked through the formalized unity/plurality relation of numbers, to achieve quantification. The generalization is internally complicated. All future doings of that object necessarily mobilize the boundaries and separations of the material arrangements of the microworld in which it clots. In contrast to Tony and Wendy, what Lucy settles on has no internal complicatedness. Lucy fails to get the separations and connections of the ordered/ordering microworld into the object she proposes as a generalization. Lucy's generalization fails as a stabilizing transition.

Effecting a stabilizing transition is dealing with complexity to effect a definition. An object clots when the repetitions and routines of its generating microworld become a ritual. The definition is more or less complicated. "A volume" is more complicated than "a thing." You must *do* more when you work with a volume; your eyes must be better trained to see as Wendy's do. I am suggesting that generalization is doing a ritual in which complexity is managed by being shifted into complicated objects. The transition effects a definitiveness. An object like a thing or a volume stabilizes and solidifies so that collective acting goes on, and so does an *eye*. There are objects whose rituals are so embedded in a going-on that it is easy to forget that they are complicated, and that their complicatedness is necessarily an outcome. The rituals of doing "things" are so routine that they are accomplished without seeming to be. So used are we to going on with them that we are prone to make a metaphysical mistake; taking an understandable shortcut, we take objects like things as givens. Seeing that some young children fail to make the generalization that has a collection of peanuts as a thing or a volume reminds us of the ritual, its microworld, and the accomplished nature of thingness.

Children with the Capacity to Generalize as Outcome

How can we understand the capacity of Tony and Wendy to generalize correctly as an outcome of collective acting in much the same way as a generalization is outcome? Participants that generalize are not givens, but like the objects I talked of in the previous section, they are outcomes of the workings of the rituals within ordered/ordering microworlds that effect generalization.

Many of the children I talked to in Melbourne, Australia, and Ile-Ife, Nigeria, showed a capacity to generalize. We might say that these children, Wendy and 'Bola for example, present as embodied with a capacity to generalize. Some younger children in both Melbourne and Ile-Ife were not enacted as generalizers. As participants in the collective going-on of their times and places, their embodiment was not yet endowed with the necessary bodily habits. We might say that they had not yet been inducted into the collective memory. When I talk of "collective memory," I picture the ordered/ordering microworlds and their rituals with which all times and places abound. I am not picturing minds!

To further explore how Wendy and 'Bola are collectively enacted as having the capacity to generalize, I turn to Paul Connerton's notion of "incorporating practices."[2] In suggesting collective memory as sedimented in bodies, Connerton points to a type of practice of which disciplined bodies are the outcome. Perhaps "worn-in" is a better term than "discipline" here, because this is not the disciplining of a human authority, which is the way we usually take the notion of disciplining bodies. Incorporating practices, routine actions, "may be highly structured and completely predictable, even though it is neither verbalised nor consciously taught and may be so automatic that it is not even recognised as isolatable pieces of behaviour. The presence of living models, of men and women actually [doing it] 'correctly' is essential for the communication in question." The effect of this is to make human bodies repositories of collective memory—the workings of microworlds in particular places and times: "[O]ur bodies . . . keep the past in an entirely effective form in their continuing ability to perform certain skilled actions. . . . In habitual memory the past is, as it were, sedimented in the body."[3]

In contrast to Wendy and 'Bola, Lucy and 'Dupe can be understood as not yet having sedimented in their bodies the necessary routines of acting, which effect them as knowing how to generalize. Nevertheless, the "wearing-in" of their bodies under the instruction of humans and nonhumans of the collective, organized as ritualizing microworlds, will, however, eventually have them so. Their bodies are still being formed within the material/symbolic practices of their times/places. They are being formed in ways that will enable them to become embodied as generalizing participants.

It is notable that these sorts of routine action (riding a bicycle and handwriting are two examples Connerton gives) cannot be well accomplished without a diminution of the conscious attention being paid to them. Each constituent event precipitates a successor without reference to conscious will. We have chains of actions in a routine accompanied by sensations, but sensations to which we are normally inattentive, and to which we must be inattentive to achieve fine performance.

Like Connerton, I am insisting on the performativeness of this sort of bodily sedimented collective memory, insisting that what is produced is not a sign. This performance cannot be interpreted in terms of some linguistic model of meaning that treats the performance as sign, an inscription that is "read." My notion of performing or enacting does not accept the notion of the body as some highly adaptable vehicle for the expression of signs of "internal cognitive categories."

The training regime from which generalizers emerge must work on habituation and incorporation of "seeings." It works not with feelings in legs and balance, as in learning to ride as bicycle, but with eyes and learning to look. Being hailed as generalizer in a little ritual with water or peanuts takes much practice in "seeing."[4]

Under the influence of foundationist tellings of realness, "seeing" has come to be understood as "observation." Rendering seeing as the perfect instrument for foundationism's removed, judging observer, the need for learning to see is often overlooked. Young children learning to walk will at first walk off a veranda edge. Soon they learn to see the drop and might become excessively fearful. In science lessons, children learn to see through microscopes. It takes weeks of practice to see a cell. The eyes of cytologists are further trained. Learning to see ultrasound images is notoriously difficult. Watching television trains children's eyes to see as the camera's eye. All these are examples of learning how to figure images, how to background and foreground.

Where do the figures by which these images make sense come from? In the last case, instruction for seeing is often carried in the narrative in which images are embedded. Watching *Sesame Street* is education in many areas of modern life, one of them being how to see the TV screen. Some episodes also explicitly teach young viewers to see generalizations, like length or volume. Instruction in how to see a microscope field is provided by encouraging teachers, urging patience and persistence in looking. Diagrams on the blackboard and in textbooks foreground and background giving the learning eye guidance. As I taught my Yoruba students how to teach children to measure, I spent a great deal of time training their eyes to see length as continuous extension. Where they might otherwise see only the edge of the laboratory floor, with string, or with the laying again and again of a long thin ruler, I insisted that they foreground a simple and uniform extension.

In the conversational doings that featured in my experiment, habituated eyes, the outcome of a regime of practice in ordered/ordering microworlds, figured water or peanuts as unitary. Learning to figure, habituating their bodies in routines of seeing and doing, has children generalizing. They come to share their worlds with "the peanuts" through generalizations like thingness, or volume, or unicity. Once they have learned to do this, it is very difficult to imagine that they could once not do it. In the same way, once we have learned to swim or ride a bike, it is difficult for us to imagine not being able to do it.

Foundationism has properties like thingness, length, and volume as things in the given world, waiting there for observers to cognize. Of course, once they have learned to do so, knowers can be certain of their generalizations about material things, using these found abstract objects. Seeing generalization in an alternative way, we have a quite new and different explanation of the certainty that generalizers feel about their generalizations. We see now that their certainty is an embodied certainty. It does not lie in their having made exhaustive observations of instances of the material world and having, on that evidence, come to a conclusion. That unlikely situation, however, is what lies inside my foundationist taking of the children's certainty as self-evident. This new way of understanding what children have achieved as they emerge as generalizers is credible. It credits children with the capacity to image, to figure, a capacity taken on under instruction from the humans and nonhuman participants of their times and places. Children have had their looking trained by those (humans and nonhumans) about them.

These contrasting origins of certainty help me realize that my former explanation of what children have achieved as they learn to generalize through learning to see some things that are not in the material world—"abstract objects"—is quite incredible. Foundationism suggests that, by learning to "see through" the confusing surface appearances of materiality, children suddenly penetrate its inner secrets. Quite improbably, I had children as learning to cognize the given categories of the real world: matter, space, and time. Admittedly, my relativist version, where the real structure is accessed through knowing the rules of language taken on under demonstrative instruction from other users, is not quite as incredible as the universalist version of children emerging as generalizers. There, children learning to conserve properties are seen as developing minds emerging into a state of grace where they suddenly access the rules of the world through observation.

The ritualized working together of water, peanuts, hands, eyes, words, bottles, cups, and glasses in quantifying is an ordered/ordering microworld. To be embodied in that microworld is to be subject to it, lined up by it. The working of its ritual effects some that we want to call subjects—the generalizing children—and some that we want to call objects—things and volume in English and in Yoruba ẹyọ (ohun) or ẹyọ (ìdì) (where I translate ẹyọ as unicity). These accomplished objects, or, we could say, this accomplishment of the unitary, were what children talked of and pointed to in their responses to my questions.

Getting the Unitary in Generalizing as an English Speaker

In my report of the experiment, I presented Lucy and Wendy as representative of monolingual English-speaking children who, respectively, are not and are embodied as generalizers. Comparing and contrasting Lucy's and Wendy's responses, we can recognize the foregrounding being effected in the episode and learn something of the difficulties in accomplishing the seeing prescribed in scientific quantifying microworlds.

Lucy has previously agreed that two tumblers of water are the same. When the water in one glass is divided equally between two smaller tumblers, she asserts, "Now there's more water in that one [the remaining original tumbler] because it's very tall and these two are low." Lucy tells me she is trying to "keep tabs on the stuff," and she does it by foregrounding one of its dimensions. She is foregrounding from the multiple water presentations generated in that episode.

Wendy has watched me twice fill an empty condensed-milk tin with peanuts. First I tip the tin of nuts into a tall teacup. The next time I empty the tin into a wide plastic bowl. One of the ways that Wendy accounted her certainty that the large flat bowl and the teacup were the same was this: "When I think of the space inside the bowl, I can see that, if I squash that [indicating with her hands the diameter of the surface of the peanuts in the bowl], it will make it higher and I can see that it's the same as in the cup." Wendy's foregrounding is not so different from Lucy's, but what she describes is more complicated. Her foregrounding is a multidimensional playing-off of boundary and what it encloses. A bounded collection of peanuts, one dimension against others, and Wendy images the unitary. Wendy's foreground is a dynamic and flexible treatment of boundary/enclosure, a more highly developed seeing than Lucy's.

What the responses have in common is that both children are advising me that you need to watch for the *stuff* to see what it is doing; you need to watch the boundary and what it encloses. Lucy is on the right track. Both Lucy and Wendy say that you have to get a fix on and learn to see a particular something about the stuff. As English speakers say (and as I said in my old report), you must get onto some quality "in" the stuff when you are learning to measure in English. A quality in the stuff is *done* by foregrounding a unit contrived by contrasting boundary and what it encloses. In this case, Wendy's habituated eyes effect the unitary. There are, of course, many other ways to effect a unit and do a quality.

Listening attentively to Wendy, we learn that, in contexts that she as an English speaker is familiar with, you need to watch boundaries and what is inside them. The problems here are getting something that is doable with the resources available—habituated eyes and hands. You need to settle on something that circumstances allow—get stuff manageably bounded. What Lucy and Wendy are learning in school is important here. Teachers are helping to train their eyes to see the images that they need to order the episode. Children need to be sensitized with story and image to get onto this or that quality in the stuff.

A child might opt for a complicated relation of boundary and what is inside to get a trackable unit. Then English speakers say the child can generalize the stuff as volume; the child can *do* volume by using the complicated unit. At the same time, a simpler focus on just boundary might suffice. Like Tony, the child then generalizes to effect a thing.

Summing up: In my new framing, I paraphrase what the English-speaking children said as "You must work from the situation and settle on a trackable unit of stuff." There were several possible options in the episodes I presented to the children. Stuff with boundary (things) is easiest. Learning how to deal with the differing sorts of trackable units is what these children learn in school as they frequently handle and deal with standard measures and hear stories of a wide variety of attributes or qualities. Justifying the enterprise of their being trained to recognize the units of doing qualities is a set of ideals expressed as images and narratives. These are the abstract qualities that the mathematics and science curricula identify as lying at the center of their project.

What this reading implies is that the young English speakers were being trained to work with some images they had been sensitized to in order to contrive a workable unit. We could sloganize it as "Get a unit and you can go on to how many and define with a plurality." It leads to doing a one/many relation. There is the added requirement that children need to listen to the stories of the ideals in order to get onto different types of units.

Two issues have emerged from paying close attention to what the English-speaking children told me about their generalizing and refraining from translating it through the figure matter/space/time. First, we see again the figuring I identified in chapter 5 as English language numbering sequence—the workings of a one/many relation. Second, the issue of the role that ideal figures play here is interesting. It seems that a fully fledged one/many doing of unity/plurality, that is, using units with boundary/enclosure relations more complicated than things, emerges along with formalized mathematics and science lessons in schools. In the ordered/ordering microworlds of classrooms, ideals come to life in telling stories and acting them out. These are the familiar math lessons of the primary school.

It seems that qualities are neither in the world nor in the culture, as universalist and relativist versions of foundationist stories would have us believe. They are in the doing of science, in ordered/ordering microworlds—laboratories and field sites, or in this case in the doing of science lessons.

Getting the Unitary in Generalizing as a Yoruba Speaker

From the responses that 'Dupe and 'Bola made, we get a quite different story about what you need to learn to foreground in generalizing. They show us another version of embedding the ordered/ordering microworld in a quantifying generalization. They also tell us what you need to keep tabs on and how you do it, but it is quite a different set of instructions. Little 'Dupe has watched while I even up the water level in two large glasses. Taking one glass, I pour it into two smaller glasses, trying to ensure an equal amount in each small glass. Having watched my precise and careful actions, she helpfully explains, "This one is a one and this one is a two" ("Ó pò nínú èyìí torí pé méjì ni wón sùgbón òkan èyìí," i.e., "There is more water in this because this is two and this is one"). 'Dupe is foregrounding an action, the pouring into two. She suggests that what needs to be watched out for is the "doing to" the water. 'Dupe's mistake is that she does not realize the correct "doing to" to give precedence to in her foregrounding. She incorrectly chooses the "doing to" of the last action, yet it is the first action that is significant here—the action of pouring water into the two large glasses and evening them up is what makes a whole, and that's what you need to look for when you are focusing on the boundary making. The second action of pouring into two smaller glasses makes parts out of that whole.

'Bola has watched as full tins of *ẹ̀pà* (peanuts) are emptied into a plastic bowl and a cup. When I ask if there are the same amounts of peanuts in the bowl and the cup, she replies, "Òkan wà níbí òkan wà lóhùn-ún" (There is one here and one there). Explaining further, she says, "Ẹyọ kan ni ẹyọ kan, àfi tí a bá pin in si méjì bẹ́ẹ̀ ni mo ni wò ọ́okò pin in" (One is one unless you divide it into two, and I watched and you didn't divide it). I asked her whether she would know they were the same if she had turned her back while I poured the peanuts out. "No," she said, "you might have divided it and taken some away to sell to another person" (Rárá ó seése ki o ti pin in ki o si ti mú díè lo tà fún elòmiràn). When in conclusion I ask 'Bola if the peanuts in the bowl and the cup look the same amount, she replies that they are not equal in that respect: "Wọn kò dógba." When I repeat the question, putting emphasis on amount,

'Bola asserts that you cannot know whether it is the same amount by looking: "Ó nira láti mò bóyá iye kan náà ni wón nipa wíwò."

'Bola specifically disagrees with Wendy over what should be looked for in a quantifying microworld. Foregrounding actions, 'Bola is setting the second partitioning act of the episode as occurring within the context created in the first. Both 'Bola and 'Dupe are saying you need to keep tabs on what is being *done to* the stuff. It is the partitioning that is important. 'Bola tells us that you must foreground actions in the right sequence, you must get what is being made as unitary before you consider its parts.

I am taking 'Bola's exhortation that we must watch what is being done to the stuff as instruction to keep eyes on boundary making, how and where boundaries are being formed and re-formed. When you are watching boundary formation, it is important to recognize the definitive event of the particular context. The first action, filling up the tin of nuts, is the definitive action of the context; the subsequent action of tipping it out is irrelevant. As 'Bola advises us, what you need to contrast is those actions that define the whole we are dealing with here, and set it in relation to those that make parts.

The first ordering act in this quantifying microworld is the definitive one—it makes the whole being dealt with. When you are routinely hailed by the definitive boundary-making act, then you can learn to work from the situation to settle on a workable way to manage it. It may be a ęyọ (ohun) unit, something that arrives as such in the context, or an ęyọ (ìdì) unit, something contrived, cobbled together. Contingency in the ordering comes after the definitive moment. The definitive act is "settling on the whole." In the subsequent contingent move, how that whole is sectioned into parts is worked out.

Here definition grows out of the quantifying context; ideal images are not necessarily involved, although presumably the act that defines the whole is itself an expression of the collective going-on of which the quantifying microworld is part. This paraphrasing of statements made by 'Bola and 'Dupe suggests that the bodies of young Yoruba speakers were being trained in the working rituals of quantifying microworlds in ways that were different from the "training regimes" young English speakers undergo. This has 'Bola and 'Dupe telling us that their eyes were being trained to light upon the defining action, the boundary making within the ritual, which was defining the unitary. Focusing on the boundary making carries the implication of doing the unitary/plurality relation as whole/part. Again we see emerging the contrast I made in chapter 5 between the ways that English numbering and Yoruba numbering do the unity/plurality relation.

Generalization: Transition from Unmanageable Complexity to Manageable Complicated Objects

I have come up with a generalization about generalization: (1) Generalization is the outcome of working a ritualized ordered/ordering microworld. We can understand a working of a particular microworld as a set of strategies for managing complexity. (2) Those who would manage complexity are made as generalizers in the ritualized workings of microworlds along with the generalizations (note: this implies that not only humans are generalizers). (3) Generalizations are complicated objects, contextually embedded accomplished rituals. In the subsequent doings of particular generalizations, the management strategies of the ritualizing ordered/ordering microworld are reenacted. (4) The ordered/ordering microworlds made in the several times and places I did my experiment were quantifying (generative of quanta). The rituals enacted in them were various strategies for effecting a defined unit from many and varied presentations. Here complexity was rendered manageable as a unit, which might in a further step be worked as the unity/plurality relation. (5) Tracking in quantifying microworlds generates logical orders through generating the unity/plurality relation, but which order emerges is underdetermined by the material order of the microworld. It is form-of-ritual-in-microworld that determines the particular way unity/plurality is done. (6) Ritualizing unity/plurality one way renders a generalizing logic of whole/part; doing it another renders a logic of one/many. On this account, differences in generalizing lie in different renderings of ritualized quantifying microworlds as the unity/plurality relation. (7) What follows from points 1–6 is the possibility of multiplicity in quantifying generalizations, in turn pointing to the possibility of many strategies for managing the multiplicity (complexity sneaks back in). (8) By recognizing alternative sequences of definition and contingency in the working of quantifying microworlds, we recognize that doing one/many is not the same as doing whole/part. In the first, ideal images shape a contingent approach to eventual definition, and in the second, definition arising from contextualized doing leads to contingency. This helps us recognize that various doings of the unity/plurality relation effect, and are effected in, moral orders.

The Generalizing of My Experiment

When I planned and carried out my experiment, I understood its objective was to account differences in generalizing. In writing a report of that experiment, I failed to account that difference in a way that was useable. Difference is accountable only when there is a general account of generalizing within which difference is feasible. The account of what all generalizing shares—the sameness of Yoruba and English generalizing—is constitutive of how they might be told as different. As we saw in chapter 7, a foundationist account of what all quantifying generalizations share cannot allow difference. The foundationist generalization about quantifying generalizations has them as ideals—pure, essential, and given as a foundation. There is no place for difference in that. There is also no way to recognize quantifying generalizing as effecting moral orders. In contrast, my new generalization about quantifying generalization enabled a telling of difference as doable and manageable. It also allows recognition of quantifying generalizing as making moral orders.

Since my general telling of generalizing is itself a generalization, necessarily offering a particular rendering of difference between logical and moral orders in quantifying, I turn now to ask about how my general telling was constituted. I am asking how my account of what all quantifying generalizations have in common was an outcome of my experiment. My generalization about generalization, my transitional stabilization, needs to be explained in terms of the ritualizing microworld from which it arose, just as I explained the children's generalizations in terms of their constitutive ritualizing microworld. We see how my generalization of generalization was emergent in just the way the children's generalizations were emergent. My report of the experiment is an account of what I did in generalizing about generalizing, just as the children gave me a report of what they did in generalizing about stuff.

The children sat with water, peanuts, bottles, cups, glasses, and so on. They captured glimpses of boundaries and stuff with their variously habituated eyes. In response to my questions, they used words they were familiar with in articulating the orderings they achieved through those glimpses and how they disciplined them in seeing. I captured those articulations. My capturing was much more complicated and messy than the children's efficient capturings with their "worn-in" eyes and familiar words. I needed a tape recorder, batteries, cassette tape, transcribers, translators, paper, pens, and index cards. Not only did my captured articulations require many more resources, trained with a great deal more effort than eyes, it also took much longer to effect their capturing. I needed hours, weeks, and months to accomplish the index cards I described in chapter 7.

Just as the children's generalizations emerged from an ordered/ordering microworld, so did mine. Just as theirs rendered the tension/resolution of boundaries that were generated within the ritual working of the microworld as an internally complicated object, so did mine. Internal to the children's generalizations were connections/separations of stuff and its boundaries. My generalization connected/separated stuff, children, and languages, making a complicated set of boundaries/enclosures. The complicated internal structure of my generalization can be understood as a clot of the orderings contrived with my little army of human and nonhuman helpers. It is a stabilizing and solidifying entity: a general account of generalizing.

The generalization of generalization is also doing politics—ontological politics. This generalization of generalization has a particular internal complicatedness. It arrays children and the languages they speak in a specific set of relations that purports to explain all manner of things: materialities and their differences, difference in generalizing between children and between children using different languages. It also explains how children and languages interact in a general way. In explaining children and languages generally, it intervenes in making them objects/subjects of particular sorts. If the generalization about generalizing that emerges from my ordered/ordering microworld is done, say, by teachers picking up on these notions and trying them out in the classroom, the order of the microworld of my experiment must be partially at least recreated—or else, the generalization will not emerge; it will not hold.

The first story in chapter 1 tells of my disconcertment in a science lesson where I experienced, as quite unmanageable, some tensions between what I now call competing ways of ritualizing in quantifying microworlds. Mr. Ojo managed that complexity by coming up with a new version of a quantifying generalization that was neither and both Yoruba and English. The new object thrived for a short time, and then died off. It may continue to emerge here and there in Yoruba classrooms, doing its work as hybrid, but it did not make it into curriculum documents, for example. As a quantifying generalization, it did not become part of my curriculum for teaching science teachers.

My response to the tensions around that complexity was similar; I also made a new ritual—an experiment. I have now shown that that ritualizing microworld can be rendered in two ways. My experimental microworld was, like that of the children's, underdetermined by the material order in which it was effected. We saw that the ritualizing microworld the children talked of could be expressed in alternative ways, either through focus on the boundary making or focus on boundaries and what they enclose. So too the microworld of my experiment does not determine the forms of the generalization of generalizing that emerge.

The microworld might be shifted down into the object "generalization" of generalization" as abstract, idealized by rendering it through the figure matter/space/time set within the tripartite modern constitution. It can equally be shifted into the object "generalization of generalization" as internally complicated, relational material object through the figure unity/plurality set in worlds as emergent. Each version of "generalization of generalization" equally embeds the ordering of the experimental microworld from which it emerged. The difference between them is that the version of "generalization of generalization" emerging in this second account, which sees generalization as transitional figuring, openly acknowledges its origins as generated in managing complexity. The other, having itself as a given ideal, denies the possibility of multiplicity and hence difference in generalization. It refuses the possibility of ontological politics.

There is something else of interest, too. We can understand the way of telling my microworld's ritual, which I elaborated in chapter 6, as one consistent with a focus on the boundary and what it encloses: the complex clot of children, stuff, and languages that I dealt with is unitized through the figure of matter set in space and time. I said I did what the English-speaking children told me they did.

Telling generalizing as transitional figuring effects a focus on the boundary *making* in the microworld. It understands itself as needing first to come up with a definition of the whole from the context. Telling generalizing in this frame of emergent worlds turns out to be analogous to a Yoruba telling: give a generalized account of generalizing before telling its parts—English and Yoruba generalizing. We see a continuity between what 'Bola says she did and what I say I did, in this second telling.

This does not necessarily imply that the explanatory frame is common. It does not necessitate that Yoruba people understand their world as emergent. They might, but that is not the issue here. What we might accept is that this way of explaining generalizing—focusing on telling the ordering in the microworlds it requires, is a more flexible explanation of generalizing. It avoids engaging particular ideals, and frees us from the highly connected modern frame that engaging the figure matter/space/time brings with it. It also confronts us with being responsible. Promoting the idea that scientists should try to emulate 'Bola in telling how they came up with their generalizations would have them telling what they did without hiding behind the tired and worn-out myth that they were just looking for something that was already there.

Part Four: Certainty

Chapter Nine: Two Consistent Logics of Numbering

This chapter begins with an argument for disparate methods of predicating-designating in English and Yoruba. I argue that difference here is significant in causing the different logical forms of quantifying in Yoruba and English. This paper was the climax of my relativist study of logics suggesting the possibility of contesting certainties located in alternative linked systems of predicating-designating, quantifying generalization, and numbering.

Chapter Ten: Decomposing Predicating-Designating as Representing

In a move familiar from chapters 4 and 7, I decompose my argument over predicating-designating categories. I show that self-reference secures my definitive contrast of English and Yoruba predicating-designating categories. Recognizing this, we see that foundationism, an explanatory framing that proscribes literalizing, having it as an impermissible category mistake, is actually secured through literalizing. The alternative chains of generalizing logic through which contesting certainties might be achieved dissolve along with the difference I purport to have discovered in chapter 9.

Chapter Eleven: Embodied Certainty and Predicating-Designating

If predicating-designating is not the first step of logically representing the world, what is it? I give a new account of predicating-designating, showing it as specific figuring of relationalities in acts and expressing the embodied certainty of collective acting. Recognizing this changes our understandings of objects generated in talk, and we see how the objects generated in Yoruba and English predicating-designating differ. This enables possibilities for articulating the ontological politics implicit in differences in generalizing logics.

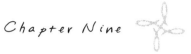

Chapter Nine

Two Consistent Logics of Numbering

*I*n cartoon style, I have been contrasting two versions of foundationism that I style universalism and relativism. Universalism has a naturally given foundation, which is represented in logic and in knowledge generated according to that logic. In contrast, relativism takes it that logic's foundation was given in past social acts that, in inventing ways of managing physical reality, laid the foundation for representing it. Universalists have certainty originating in the proper correspondence between the categories of logic and the world (or the experience of it)—a correspondence theory of truth. Relativists adhere to a coherence theory of truth, where certainty is guaranteed by consistency in a system of representation. What matters in the system is that the meanings words carry with them seamlessly link up with other words.

This chapter is a paper arguing this relativist position by showing alternative internally linked systems of generalizing logic in English language (and by extension science) and in Yoruba language.[1] Published in 1990 in what is perhaps science studies' most prestigious journal, the paper repeats a little of what I have already presented in chapters 3 and 6. Since this paper is a final synthesis of my project of demolishing the universalist account of natural number, I present the paper in its entirety and ask my readers to bear with the redundancy.

Investigating the Social Foundations of Mathematics: Natural Number in Culturally Diverse Forms of Life

How can we expose the social foundations of mathematics? Natural number seems a good place to start, for it is among the most sacred of the Platonists' objects; it is one of the least problematic of the empiricist's reality; and for the psychologist, it is the most obvious of mental structures. In this paper, I reveal natural number as a social product by demonstrating its origins in both talk and material practices. We use natural number without wondering how and why it works. We grasp the rules by which natural number is constituted without ever bothering to consider how they might be discursively elaborated. The certainty associated with number use is such that it remains quite resistant to study as a social phenomenon. What is behind the work that natural number does for us? What constitutes its working? I suggest that there are three sets of rules to be grasped in coming to number, not discursively, but practically. These are rules of practice, rules we acquire through training. And having learned how to use the objects constructed through their use, we just go on, taking these objects, numbers, as self-evidently "facts of nature."

The ordinary user may have little idea of where to begin elaborating the rules by which number is generated, but this does not imply it cannot be done. We can reveal natural number as a social product by means of demonstrating its origins in both talk and material practices—its constitution in social practices.[2] In what follows, I elaborate what underlies our use of number names in talk by showing the series of generalizations by which natural number is constituted through practice in two communities that are culturally disparate: the Yoruba community of West Africa and the English-speaking community.

A View of Language: The Practice of Referring

Laying bare the methodological nature of natural number is to unpack the ways that the practice of referring in ordinary language use, working together with ordinary material practices, have come to constitute the quite precise meanings that number names carry with them when used in talk (and writing). So, as a prelude to "telling the story" of natural number as founded in social practice, I need an account of the linguistic practice of referring in the sense of designating, in making meaningful sentences. This is an attempt to account for the working of language through detailing the social practices in which the code that is language develops. To look at the origins of language in this way is to study the practices of very young children working with the competent language users around them, as they learn to use the code of language and so learn to designate and to predicate something of what it is they designate.

In showing how natural number is constituted through use of language and practical methods, I need a view of language that serves my purpose. More particularly, I need an account of reference as social practice. An account of language that traces the origins of making meaning to the referent as a "real" object, as linguists of a functionalist persuasion do, is not useful. No more useful is the structuralist account that has language as a system of differences created within a signing code. I need an account of reference as an apparatus that develops from the procedures and methods that language users adopt in making meaning. But let us not be mistaken in the search for such an account: we are not looking for the essence of language. Reference is not the essence of language—there is no essence of language.

Language is a process of encoding. In common with all the signing systems used in making meaning, language starts with the inductive generalization of the permanence of bodies. Babies have already got this far when they smile at their mothers and no one and nothing else. In developing use of the signing code language, classifying is central. There are two distinct types of classifying that beginning talkers engage in: classifying types of bodies, and classifying types of interactions between bodies. The first of these classificatory acts has children accumulating lexical items that come to function as nouns, general terms. In English, these are terms like "mama" and "doggy." The boundaries of the classes labeled with such terms are of various sorts; children learn the boundaries and the classificatory labels by ostensive training.

We often think that it is through extending their use of this sort of classifying that children come to talk of the world "properly," that is, learn to refer or designate. However, classifying bodies is only secondarily involved in learning to refer/designate. The real strength of the encoding practices that come to be useful in making meaning lies in encoding the actions that bodies engage in. This second classification results in children's accumulating lexical items that in time will come to function as predicators (verbs) in making meaningful sentences. Children growing up in English-speaking communities say "down" (as they gleefully drop their crust for the fifth time and wait expectantly for it to be "upped"), or as they thrust away their bowl, "orgon!" Learning the particular way to make this sort of classification and label it is also accomplished under ostensive training. It is the beginning of learning the signing activity of predicating, learning about using verbs. Learning to predicate (use verbs) is an accomplishment of great importance for every language learner, for along with predicating comes referring or designating.

Designating and Predicating

Learning to mean with language starts with classifying, but classifying does not amount to making meaningful sentences. Meaningful sentences are uttered within the framework of a particular "vision" of the world, a "vision" that the language used in making meaning necessarily and inevitably encodes. Even those apparently transparent acts of referring, observation sentences, where designating categories are used in stimulus situations (as in "This is black and white"), come from within this "vision" of the world. The designant or referent is not the actual object. The designant is a certain type of category.

To say that sentences carry meaning is to recognize the importance of predication in meaning. Predication develops from the more complex of the classifying activities in what we might call protolanguage use: classifying actions of bodies. It is these classifiers that come to act as verbs. In incorporating the role of predicator into use of the code (using verb terms), the language user necessarily incorporates a particular sort of classification of difference, and is thus pushed into identifying and specifying the subject of the sentence as a particular sort of category.

In this, there has been a leap from protolanguage to language, from making classifications to making meaningful sentences. A "vision" of the world has necessarily been engaged through adopting a particular way of classifying actions of bodies in coming to use verbs, and the sorts of things that constitute the world are inevitably postulated. A specific sort of referring or designating category has been adopted. This particular type of referring category becomes an obligatory assumption of all those who use the language. It is a postulate about the type of the constituents of the material world. While this category is inscrutable, we have no way of telling how it fits with the "real" world; it is not ineffable.[3] We can identify and articulate what sort of a category it is by working from our understanding of the rules we use in using verbs, the criteria of our classification of actions of bodies, from which verbs grew.

This view of designation differs from that offered by linguists of the structuralist persuasion, who, while they concede the existence of desig-nating categories, have such categories as ineffable.[4] Those linguists work-ing with a functionalist approach are sensitive to the role that training in classification plays in learning to talk. However, they generally conflate referring category and object.[5]

Such an account of designating in language use is an articulation of the rules that lie behind the use of the categories of a language; these are the rules we teach our children by example and through negotiating with them over what is allowed and what is not while they are learning to talk. Articulating the rules in this way enables description of the type of designating category that use of a particular language commonly entails. The type that a designating category is, typically, is of no interest to a user of that language. Consideration of its type is superfluous as we use it through signing with it. As language learners, we labored to get it as we learned to use verbs in all manner of material and social contexts, and since we got it, we have taken it for granted. Language use is an example, par excellence, of grasping rules in a way that is not interpretation.

Talking about the World in English and in Yoruba

I go on now to compare the types of objects postulated as constituents of the material world by English speakers and Yoruba speakers as they engage in ordinary talk of the world. It is an initial step in coming to number. I suggest that the types of things that English speakers postulate as constituents of the world differ from the types of things that Yoruba speakers have as constituting the world; that is, the types of designating categories used in the two languages differ. I go on to elaborate the conse-quences this difference has in the constitution of natural number in those two communities. To use predicators in talk precipitates use of referring categories in talk, and in using referring or designating categories we are accepting postulates about the types of material objects that are in the world. The postulates are built into our language use, and we rarely be-come aware of them. In what follows, I describe the types of designating categories used in sentences that might be uttered in a stimulus situation using English and using Yoruba.

I suggest that there is a significant difference in response to the English question "What is it?" and the Yoruba question "Kí nìyìí?" although the questions and answers point to the same manifestation. The difference between the responses is a difference in the type of thing taken to be in the world. My linguistic experiment explores the methods of working of verbs (predicating), of subjects (designating), and of objects (predicates), in English and Yoruba sentences. It is an exercise very similar to what in my primary school English lessons was called parsing. In this parsing exercise, I ask you to imagine something in front of you about which you are curious. If you are an English speaker, you might ask your companion, "What is this?" If you are a Yoruba speaker, you would ask, "Kí nìyìí?" Presuming your friend is able and disposed to satisfy your curiosity, she will answer you by designating and predicating. These processes are the routine workhorses of language use, and in them logics begin to condense. I am going to look at the nature of this in English and Yoruba by analyzing two possible responses to "What is this?" and "Kí nìyìí?"

It is easy to imagine a situation where an English-speaking informant will tell her curious companion who has asked "What is this?" "It is a hoe." Or she might say "A hoe." In a different situation, the answer might be "It is water" or "Water." In the first situation, a Yoruba-speaking informant would tell her friend who has asked "Kí nìyìí?" "Ọkọ́ ni ó jẹ́" or "Ọkọ́ ni." In the second situation, the reply would be "Omi ni ó jẹ́" or "Omi ni."[6] In what follows, I investigate the working of the verbs *is* and *jẹ́* to see the foregrounding/backgrounding implicit in their presenting of acting (in this case the acting of manifesting or being) and what follows from this.

In Answer to "What Is This?"

"It is a hoe" and "It is water" are simple utterances in English. They comprise three basic elements: the predicating term "is," the designating term "it," and the predicate expression containing characterizing terms "a hoe" and "water," ascribing or attributing particular characteristics to what is designated. We can explore the foregrounding/backgrounding in predicating-designating in two ways—by looking at the workings of verbs and by paraphrasing designating terms, albeit into awkward forms.

A speaker who announces "It is a hoe" may be taken as meaning "The thing here and now is a hoe." Similarly, in saying "It is water," a person could be taken as meaning "The thing here and now is (a glass of) water." "It" evokes a here-and-nowness. We may paraphrase the designating term "it" as "this spatiotemporal particular." The speaker is identifying a section of the material world that is here-and-now. Its here-and-nowness is what defines it as a particular. "It" is a separated section of matter that extends in space and endures through time.

Where does the specification that achieves this particularization come from? There is nothing in the word "it," and nothing in the dirty metal and wood hoe or the clear enclosed liquid that can determine this specification. There are no resources somehow buried in the "it" or in the hoe that determines that "it" must evoke a spatiotemporal particular. The specifications derive from the coding action embedded in the working of the predicator "is." Predicating is acting to foreground and background in a particular way in presenting acting, in this case the acting of being or manifesting.

Look at verbs in English—what characterizes their working? They take on differing forms that show tense: past, present, or future forms of a verb must be specified in speaking English. Specification of time is mandatory. They are either transitive or intransitive: they specify, completely or incompletely, the spatial characteristics of an act. Those are the mechanics of verbs in English. The mechanics of their working indicate that they foreground and specify a spatiotemporal particular. Individuated sections of matter located in space and across time is the type of particular that English speakers designate. In making this categorization, the possibility of a second-order categorization is constituted. The English speaker who says "It is a hoe" or "It is water" is in the predicate also evoking in a secondary way the kind of spatiotemporal particular: ". . . is a hoe" and ". . . is water." The spatiotemporal particular manifests a certain sort of materiality, the set of characteristics that go to make up hoeness or waterness. In English, spatiotemporal particulars designated as an outcome of predicating are further specified—qualified—in predicates.

In English, particulars individuated on the basis of spatiotemporal location, that is, as space-filling and enduring, are of two major kinds. The existence of different types of characterizing terms in predicate expressions in English derives from this categorical distinction, which in turn follows from using spatiotemporal location as a primary basis for categorization. It seems that the obligation placed on English speakers to use predicate expressions either with a phrase like ". . . is a hoe," or with a mass-characterizing term like ". . . is water," is a consequence of primarily specifying spatiotemporal particulars. Some spatiotemporal particulars feature an enduring boundary and are considered functionally integrated units—things. This boundary is used only if the enclosed spatial extension falls within a certain range relative to the capacities of human hands. Matter that is not said to be individual things may have no perceptible internal boundaries, like honey or water, or the bounded spatial extension may not fall within the range that is convenient for the human hands, like sugar and sand. In both these cases, matter is characterized as continuous—it flows.

In reply to the question "What is this?" an English speaker designates a spatiotemporal particular; a secondary categorization using characteristics of matter qualifies this particular. Within this second-order categorizing, there is a distinction relating to the type of boundary attributed to the spatiotemporal thing.

In Answer to "Kí Nìyìí?"

The subject of a sentence designates. Which term is serving this function in the Yoruba utterances "Ọkọ́ ni ó jẹ́" and "Omi ni ó jẹ́," given as replies to "Kí nìyìí?"? According to established manuals of translation, ọkọ́ and omi, nouns translated as "hoe" and "water," respectively, can be understood as designating in these sentences.[7] The word ni has no translatable meaning. It functions to introduce an element that emphasizes.[8] Such introducers occur just before the element they are introducing; the emphasizing element in the sentences above can thus be recognized as ó jẹ́. This idea of an emphasizer is not easy for an English speaker to understand, but it is an important clue to how predicating and designating work in Yoruba.

Ọkọ́ and omi name what the sentences "Ọkọ́ ni ó jẹ́" and "Omi ni ó jẹ́" are about; ọkọ́ and omi are designating terms. But what is designated? What is the category? The meaning we take from the verb will help in identifying this. So too will the introducer, which points to the verb phrase as emphasizer. Ó jẹ́ is generally translated as "it is"; however, the verb jẹ́ has a different meaning than is in English. Rather, jẹ́ implies "exists or manifests with its inherent characteristics."[9] So ni ó jẹ́ can be regarded as emphasizing the intrinsically characteristic way in which the manifestation is occurring. And the only things that can manifest in their characteristic way are those that are defined by those very characteristics; what is being foregrounded is a set of characteristics.

The categories conjured up by ọkọ́ (hoe) and omi (water) in the sentences above are defined by the criterion of having that peculiar set of characteristics that go to make up ọkọ́-matter and omi-matter. I have adopted the name "sortal particular" for this type of designating category. To translate "Ọkọ́ ni ó jẹ́" and "Omi ni ó jẹ́" as "It is a hoe" and "It is water" is misleading in an important and basic way. Better translations of these two Yoruba language utterances, in the sense that they more precisely convey the meaning, would be "Hoematter manifests" and "Watermatter manifests."

Let us examine some further examples of utterances in Yoruba to see this distinction. "Óunjẹ wà" (Food is there/foodmatter [matter with the characteristics of food] manifests there); "Òjò rò ní àná" (It rained yesterday/rainmatter [matter with the characteristics of rain] rained yesterday); "Ewúrẹ́ gọ púpó" (Goats are very stupid/goatmatter [matter with the characteristics of goat] stupids excessively).[10] In these utterances, the designating term, the noun, tells us about the sort of matter being referred to: foodmatter, rainmatter, and goatmatter.

Designating terms in Yoruba language utterances conjure up sortal particulars; spatiotemporal situation is not involved in this reference. Yet a secondary division is available to speakers of Yoruba, where the criterion is spatiotemporal situation. The sortal particular of Yoruba language talk can be modalized in the predicate in a secondary categorization. Hence in *ni ó jẹ́* (manifests in a characteristic way here and now), the here-and-nowness is given by the verb showing tense. What is achieved in the predicate is spatiotemporal situation of the sortal particular.

Talk in Yoruba is of sortal particulars; the spatiotemporal situation of matter may be irrelevant. So we should not be surprised to find that it is possible to make tenseless utterances in Yoruba. For example, "Ẹ̀wọ̀n áni 'bi ó wù ú" (The chain breaks where it pleases). Many such sentences are proverbial or aphoristic in nature. Further, Yoruba verbs may be transitive or intransitive, but transitivity here implies a differing form of "transfer" than that of English verbs. Transitivity in Yoruba verbs does not imply a completed spatiotemporally located action, but rather "transfers" in the sense of pointing to a realization of a particular state of being.

In Yoruba, nouns are modalized or modified by noun modifiers. In contrast, English nouns are "qualified"—have qualities ascribed. Also, Yoruba verbs may have their action modalized or modified by verb modifiers. For Yoruba speakers, it is possible to specify the mode of action in the future, or nonfuture (past/present), and to provide complex further details of other aspects of the action. Whether the action has a fixed beginning or a fixed ending, or whether it is a habitual action, a common action, a consecutive action, or a nonoccurring action, may be indicated by the insertion of appropriate verb modifiers.

Verbs in Yoruba foreground states of being. They may be active or passive states of being. Verb modifiers differentiate the type of action. For example, one might say, "Ó jókòó" (He is in a state of sitting down—continuous action—he is seated), and "Ó ń jókòó" (He is in the state of sitting down—past/present action—he is in the act of sitting down). Examples of verbs that indicate states of being are *jẹ́* (reporting a state of characteristic being—one of several verbs that may be translated this way), *dùn* (the state of being sweet), *wọ́n* (the state of being expensive), *nípọn* (the state of being thick), and *yi* (the state of being tough). An English speaker says, "It is sweet," "It is thick," or "It is expensive." A Yoruba speaker says, "Ó dùn" (It sweets), "Ó wọ́n" (It expensives), or "Ó nípọn" (It thicks).

"Sweet" is an adjective in English, a quality of a spatiotemporal particular. But with sortal particulars designated in Yoruba, it should come as no surprise to find that here it is a verb. "It is sweet" and "ó dùn" might equally be used of a cake in a particular situation. But they achieve this equivalent meaning in different ways. The spatiotemporal particular that the English speaker is attending to is predicated as manifesting here and now and exhibiting a certain quality (sweetness) in that manifesting. The sortal particular that the Yoruba speaker is attending to is predicated as having a sweet existence.

Because of the different foregrounding/backgrounding that lies within the functioning of *is* and *jẹ́* as predicators, an English speaker ends up evoking a spatially separated enduring thing, and a Yoruba speaker ends up with things of a certain sort. This is illustrated by the literal translations of "omi ni ó jẹ́" and "ọkọ́ ni ó jẹ́": "watermatter (matter with the characteristics of waterness) here manifests its inner, intrinsic and permanent nature" and "hoematter (matter with the characteristics of hoeness) here manifests its inner, intrinsic and permanent nature."

Spatiotemporal Particulars and Sortal Particulars

The different practices in classification that underlie the generation of predicating terms in English and Yoruba create different types of referring categories. I have proposed that referring develops from the practice of juxtaposing two types of classifiers—classifiers of types of bodies and classifiers of types of actions that bodies engage in. The type of classification wrought with respect to interaction between bodies will determine the type that the referring category is. In English, we have verbs, which code the interaction between bodies on the basis of spatially individuated bodies taken to endure across time, as the interacting element. The referring category created by the functioning of verbs in the code of English language is the category of spatiotemporal particulars. In Yoruba language, interaction between bodies is coded as one type of matter interacting with other types of matter; the interacting element is treated as a particular sort of matter. The type of referring category implicit in Yoruba language use is the category of sortal particular. The different practices in classification that underlie the generation of predicating terms in English and Yoruba create different types of referring categories. This difference is important when talk is used in association with other practices to create further concepts like natural number.

There is a strong commonsense idea that there is what there is in the nonlinguistic world. It also seems commonsensical to extend this, so that people, when talking of the world, must necessarily talk of the same types of things. My suggestion that Yoruba speakers and English speakers talk of different types of material objects defies this commonsense idea. People walk around and sit upon chairs and put their arms around people irrespective of what they may say is in the world. Possibly too, they have similar pictorial images in their minds. An English-speaking person and a Yoruba-speaking person handle a cup of water in the same way, and they might wield a hoe in a similar fashion. But, despite these similarities of physical manipulation, they talk of them in different ways, because in the practice of using their language to refer, they allocate primacy to different aspects of the matter they say is in the world.

I have described a profound difference between use of the Yoruba language and use of the English language. Why has this basic difference between English and Yoruba not previously been pointed out? There are after all several million people who are, to varying degrees, bilingual in English and Yoruba. The explanation lies in the fact that the categorical scheme inherent in a language is usually held unreflectively by speakers of that language—it is the taken-for-grantedness implicit in speaking that language. It is not common for speakers of a language to examine what types of material objects their language commits them to. Rather, the difference will be noticed as a difficulty in translation. To learn a second language is to repeat the same process as when one acquired one's first language. It is a socially mediated process through which meaning is constructed. To suggest that, by merely knowing two languages, one must be aware of different designating categories of those languages is to hold a simplistic view of language use. Nevertheless, we might ask about the effects of having two distinct ways of categorizing the world on the cognition of those who are deeply bilingual in this radical sense. It seems that this can have the effect of enhancing cognitive growth.[11]

Construing Units of Matter as a Prelude to Ascribing Value with Number

I pass now from a concern with practices associated with use of the code of language, to consider practices more directly involved with using natural number. Natural number use is tied up with the allocation of value to matter, a practice that most likely grew up in association with the exchange of goods. Ordinary people are often interested in knowing how much of something there is. They need to be able to talk about "the amount" in a way that is mutually agreeable. But before they can couch their talk in number names and talk about amount, people must engage conventions by which matter is construed as manifesting in separate units of some sort—the quanta of quantifying. In this section, I argue that the ways in which English speakers and Yoruba speakers talk about the idea of amount, and construct units of matter, is wrapped up with the way they talk about matter. Talking about matter in a particular way is a constituent practice in coming to talk about units of matter, and both these practices are constitutive of the practice of ascribing value with number names.

Using number names in talk to report some notion of value in the material world is reasonable.[12] Members of a language community abide by rules in incorporating number names into talk, albeit unreflectively. In this way, subjective judgments are eliminated. Using reasonable notions of value to regulate the exchange of goods reduces argument. Through numbers, people more easily come to agreement because they lift the issue of how much there is to a plane where they can more easily agree.

One can buy a number of oranges, a number of sections of rope, or a number of calabashes of palm wine or of bottles of beer. If one is interested in more expensive goods, one might buy a number of plots of land or nuggets of gold. In each of these exchanges, number is taken as varying with amount. When one buys oranges, what one expects to get is all that stuff enclosed by the several orange skins. That is, all the matter enclosed by spatial boundaries that endure over time. On buying rope, it is the distance between two ends that is of interest. If one buys a beverage, one expects to get all the liquid that fills a container, a certain three-dimensional space. If a person buys a plot of land, she expects to have allocated to her use a surface, a two-dimensional space, a space that is enclosed by a boundary that she can walk around. In buying nuggets of gold, we may (crudely) estimate the amount of stuff by feeling the force it applies to our hand. For a more precise estimate, we may use a machine to measure the downward force it exerts.

In each of these five types of transactions, the buyer, the seller, and any onlookers have a clear idea of what has been paid for. But on each occasion, people are using different ideas of what can constitute amount, and different ideas of the basic unit that counts in ascribing value. Amount in oranges is thought of in one way (the stuff inside several skins), but amount in rope is different. What counts for amount in gold differs from what counts for amount in land. For various types of transactions, people engage different sorts of percepts, which in each case are taken as varying concomitantly with "all the stuff here and now." These percepts that buyers and sellers confidently allude to are created through generalization about stimulus situations. They are created through recognition of similarities and differences across the time span of stimulus situations. The units created are reasonable, but the basis of their creation is nonlinguistic. The units are nonlinguistic concepts.

The various percepts taken to be indicators of amount in exchange transactions seem to transcend language barriers. The English-speaking visitor to Yorubaland and the Yoruba-speaking visitor to England feel a comforting familiarity while watching the foreigners ascribing value during the exchange of goods. Their actions show that they are using the familiar ways of perceiving "the amount of stuff here and now." Although the talk of foreigners cannot be understood, and the ways percepts are talked of and the things they are said to be are unknown, a stranger may still successfully participate in exchange.

To assert that there appear to be several common ways of perceiving amount, which people in different language communities use in exchange activities, does not, however, imply anything about how people will talk about amount. It says nothing about the type of entity that amount will be said to be. The ways in which these percepts enter talk will depend on the type of material object that the language used in talk commits its users to. Amount will be rendered in talk in a way that derives from the way matter is construed in talk. Percepts will be talked of as features of the material objects said to be in the world. It could be said that the features that the material objects talked of are said to have are analogues of what people perceive when they attend to stimuli that are evoked in physical involvement with matter. I am postulating concomitant variation between what there is and what there is said to be, but it does not follow from this that speakers of different languages will necessarily say there are the same types of things in the world, nor will the things that are talked of necessarily be said to have the same types of features.

It is through talk of features of things that talk of amount of matter is made possible. Now the material objects that Yoruba-speaking people talk of as being in the world are of a different type than the material objects that English speakers talk of. Different types of objects will be talked of as having different types of features. In other words, we might expect that features that things are said to have and that are used in English language quantification will differ in type from the features that things are said to have and that are used in Yoruba language process of ascribing value, or in *iye* (saying the amount of something), as Yoruba speakers would have it. Talk of the various percepts in which the idea of amount of matter resides will be talk of quite different types of features—different types of abstract objects—in English language talk and in Yoruba language talk.

Units of Qualities: English Language Quantification

The material objects that English speakers talk of are spatiotemporally defined objects of different sorts. For English speakers, the primary criterion by which the objects talked of are defined is the idea of matter set in relation to the idea of space-time. Secondarily, these objects are categorized along sortal lines. Features that spatiotemporal things are said to have name the different inherent characteristics that things of different sorts are supposed to have. English-speaking people come to say that the spatiotemporal particulars that they talk of have qualities. Qualities do not purport to name material objects. Toes cannot be kicked against qualities, nor fingers dipped into them. They are abstract objects.

The idea of amount of matter enters English language talk through talk of qualities. The extent to which some qualities are held or exhibited by spatiotemporal particulars may be taken as varying concomitantly with the total amount of matter that constitutes that spatiotemporal particular. These qualities are those that are most commonly exhibited: numerosity, volume, area, length, and weight. In ordinary English language quantification, the extent to which any of these qualities is exhibited by a spatiotemporal particular can be reported with a number. That number is taken as reporting "the total amount of matter manifesting."

Reference in English language talk is to spatiotemporally defined objects. It seems that a consequence of this is an explicit division in talk between matter that is treated as a collection of enduringly bounded integrated wholes—things (matter taken as exhibiting the quality of numerosity)—and matter that is taken to be an undivided cumulus. This explicit attribution of the quality "thingness" or numerosity creates the distinction we see in predicate expressions in English language talk between general and mass terms: "it is a hoe" and "it is water."

In coming to everyday quantification, English-speaking people necessarily talk either of collections of individual things or of cumulations of continuous matter because their language has them referring to spatio-temporal particulars. Collections of individual things are said to have the quality of numerosity to a certain extent. Cumulations of continuous matter are said to have the qualities of length, area, volume, and weight to varying extents. When numerosity is used as the basis of quantification, the process is called counting. When other qualities are used in the quantification of matter, the process is called measuring.

In counting, individual things (units of the quality of numerosity) and integers are taken to be analogous. When other qualities provide the basis for quantification, the sections of matter, temporarily individuated, relate operationally to the percept through which the quality is constructed, just as things relate operationally to the percept through which numerosity is constructed.

In measurement, the operations by which the temporarily individuated units are created vary in their precision. In the case of formal units, like units of the metric system (liters and meters), operational rules have been formalized, and the units are standardized. In the case of casual units like cups or footsteps, the operational rules for devising units are less precise, and the units are not standardized. Either way, sections of continuous matter come to be named as units of qualities, just as "things" is the name given to units of the quality of numerosity. In measuring, these named units and integers are taken as analogues.

Quantifying in English language talk is a form of second-order qualifying. In a quantification statement, it is the extent of a quality that is being reported by the numeral. In table 9.1, I have summarized the sequence of constituent practices that make up counting and measuring.

Modes as Part of Iye (Valuing) in Yoruba Language Talk

Among other things, Yoruba-speaking people buy and sell oranges, rope and beverages, plots of land and nuggets of gold. The actions they use in these exchange transactions are familiar and recognizable to English speakers, who by watching can easily render in English language terms what has been bought and sold (in terms of extents of qualities of spatio-temporal particulars). An amount of matter changes hands and people are satisfied because a reasonable notion of value is used in the transac-

Table 9.1. Summary of the Sequence of Abstractions Composing Quantification in English

Talk of material objects	Matter construed as spatiotemporal particulars of different sorts	
Talk of features of material objects (talk of qualities)	Spatiotemporal particulars construed as exhibiting various qualities to varying extents	
Talk implicitly utilizing a specific quality	Quality of thingness/numerosity constitutes the basis for an explicit division into two types of matter	
	Divided matter: collections of individual things	Undivided matter: cumulations of continuous matter
Qualitative basis of quantification	Collections construed as extensions of the quality of numerosity	Cumulations construed as extensions of the qualities length, area, capacity, and weight
	Units of numerosity: individual things	Units of length, area, capacity, and weight created from operational context (temporary boundaries imposed)
	Units of numerosity (things) taken as analogous to integers	Units of length, area, capacity, and weight taken as analogous to integers
Two operational types of quantification	Counting	Measuring

tion, but how does the idea of amount in each of these different types of matter enter Yoruba language talk, so that reasonable valuation becomes possible?

Before we get to that, let us consider how a Yoruba speaker would report the common percepts alluded to by buyers and sellers in their actions. These are the percepts I identified as entering English language talk as qualities like numerosity, length, area, and weight. When the material objects talked of are sortal particulars, these percepts are construed as states of being. They are characteristic ways of existing or manifesting, and they are reported by verbs in Yoruba language talk. Consider the simple Yoruba sentences in the following paragraphs. The sentences use ó and wọn as their subjects. These terms translate respectively as "that which is characterized by being spatially individuated" (sortal it) and "that which is characterized by being a collection" (sortal they). The use of these terms as subjects implies that a specific stimulus situation is being talked of. Here spatiotemporal situation becomes a defining characteristic of the sortal particular designated. In this situation, sortal particularity coincides with spatiotemporal particularity, resulting in a similarity between *ó* and *wọn* in Yoruba language talk and *it* and *they* in English language use.

In a specific stimulus situation, one might say of oranges "wọn pọ" (that which is characterized by being a collection of manys; they many; they are in a state of being numerous); of rope one might say "ó gùn" (that which is characterized by being individuated longs; it longs; it is in a state of being long); of a container of palm wine one might say "ó tobi" (that which is characterized by being individuated bigs; it bigs; it is in a state of being voluminous). A plot of land may be spoken of as "ó fẹ" (that which is characterized by being individuated spreads; it spreads; it is in the state of having a large surface); and of a nugget of gold, one might say "ó wúwo" (that which is characterized by being individuated heavys; it heavys; it is in a state of being heavy). In each case, the perception of how the matter under consideration manifests is reported in predicating, using a verb. It is the state of being (an action) in which the sortal particular manifests that is mentioned when it comes to talking in Yoruba of the percepts that relate to the idea of amount.

By introducing modifiers into the sentences, a sense of degree or comparison may be conveyed. One may say "wọn pọ púpọ̀" (that which is characterized by being a collection manys exceedingly; they many exceedingly; they are very numerous), or "ó gùn díẹ̀" (that which is characterized by being individuated longs little; it longs a little; it is not long), and one may say "ó tobi jù u lọ" (that which is characterized by being individuated bigs surpasses goes [beyond] that [other] which is characterized by being individuated; it is a bigger one), or "kò fẹ̀ tó" (that which is characterized by being individuated does not wide reaches that [other] which is characterized by being individuated; it is not as large in area). In this way, the idea of comparison in amount is effected. There are many verbs in Yoruba through which subtle differences in characteristic ways of manifesting can be conveyed to achieve comparison with respect to the notion of amount.

To effect comparison between two sortal particulars in situations where their sortal particularity happens to coincide with their spatiotemporal particularity, however, has limited utility. Unreasonable judgments have not been eliminated, and comparison is confined to specific stimulus situations. In contrast, use of number names does achieve both these. Number names enable general and recognizably reasonable comparisons to be made. In valuing using number names in Yoruba language utterances, one narrows the focus of talk, so that there is precision about what exactly is being talked of. A Yoruba speaker comes to talk of the type of the here/now or then/there manifestation of sortal particulars, and to do this is to abstract from sortal particulars. To talk of a specific here/now or then/there manifestation of a sortal particular is to talk of a mode. It is through talk of these abstract objects, modes, that general application and precision is gained in Yoruba language talk of amount.

The objects that Yoruba speakers are committed to saying there are in the world are sortal particulars: material objects defined through their particular nature. Certain sets of characteristics form definitive boundaries of the material objects that Yoruba speakers talk of, and the objects talked of are construed as being infinitely scattered through space and time. The here/now-ness or then/there-ness of a manifestation is irrelevant in creating an object to talk of. We could say that qualities are the boundary lines of the objects primarily talked of by Yoruba speakers. Sortal particulars cannot be ascribed qualities, since it is on the basis of a qualitative distinction that they come into being as objects to talk of.

When one speaks of sortal particulars and one wants to say more about them than merely that they manifest in their characteristic ways, it is on their *mode* of manifesting that one must comment. Mode is the manner in which a sortal particular manifests: a here/now or then/there (spatio-temporal) slice of a sortal particular. When sortal particulars are the primary material objects that one speaks of, the modes in which they may be said to manifest are features of these objects. In quantification in Yoruba language talk, it is through mode that number comes to report the idea of total amount of matter present. Valuation is a form of modification in Yoruba language talk.

The grammatical mechanism that modifies the sortal particulars that Yoruba speakers talk of in the operation of ascribing value involves the incorporation of nominalized verb phrases in an elided form. Numerals function as nouns in Yoruba language discourse, and from their form we may take it that they are abstract mode nouns, elisions of nominalized verb phrases.[13] We can say that the numerals used in talk of matter name a mode of being of a sortal particular. Thus one may say in Yoruba, "Ó fún mi ni òkúta mérin," which is conventionally translated as "He gave me four stones." A more literal translation is "He gave me stonematter in the mode of a group in the mode of four" or "He gave me matter with characteristics of stoneness manifesting here/now as a collection divided to the extent of four." Similarly, we may have "Ó rí ajá méta" (He saw three dogs), which is more precisely translated as "He saw dogmatter in the mode of a group in the mode of three," or "He saw matter with the characteristics of dogness manifesting here/now as a collection divided to the extent of three."

Sortal particulars are the primary material objects that Yoruba speakers talk of; thus in coming to notions of value, the scheme by which matter is construed as divided involves talk of the mode of the sortal particular. They are also construed as exhibiting a degree of spatiotemporal dividedness, which is a mode of the collected mode. Thus the use of number names in Yoruba language talk is a second-order modification.

For the sortal particulars that Yoruba speakers talk of to be construed as collected, there must be agreed-upon units that constitute the collection. There must be what Yoruba speakers would call *ẹyọ* (units, elements, or items). In the process of *iye* (valuing), *ẹyọ* are taken as analogous to the integers.

The elements *(ẹyọ)* whose collection can make up a mode of dividedness may be of two types. They may be *ẹyọ (ohun)*, elements enclosed by an enduring spatiotemporal boundary, or *ẹyọ (ìdì)*, elements formed by binding together or aggregating.[14] The type of element is determined by the context of the quantification procedure. For example, one may put value on mangoes using either *ẹyọ (ohun)* or *ẹyọ (ìdì)*. One may chose to take individual mango fruits, *ẹyọ (ohun)* as analogous to integers, but in another context one may chose to take full baskets of mangoes, *ẹyọ (ìdì,* an aggregated unit of mangoes), as analogous to integers. The type of unit chosen determines the name given to the quantification process. To place value using *ẹyọ (ohun)* is to *kà*. To value matter using *ẹyọ (ìdì)* is to *wọ̀n*. In everyday usage, matter like oranges, mangoes, kola nuts, yams, and cloth may be either *kà* or *wọ̀n* as one chooses. Some types of matter may only be *kà*, like people and cars. Some types of matter may only be *wọ̀n*, like land or palm wine. In the case of *ẹyọ (ohun)*, the type of unit is implied in a quantification statement, as in "Mo rí ajá mẹ́rin" (I saw five dogs; I saw dogmatter in the mode collected, in the mode five). In the case of *ẹyọ (ìdì)*, the type of unit is specified: "Ó gbé omi garawa méwàá" (He has carried ten calabashes of water; he has carried watermatter in calabashes in the mode collected in the mode ten). I have summarized the steps in coming to *kà* and *wọ̀n* in Yoruba language talk in table 9.2.

In translation, *kà* is usually rendered as "counting" and *wọ̀n* as "measuring." We can now see that this is misleading because the categorical scheme that underlies *kà* and *wọ̀n* is quite different from that which un-

Table 9.2. Summary of the Sequence of Abstractions Composing Quantification
        in Yoruba

Talk of material objects	Matter construed as sortal particulars
Talk of features of material objects (talk of modes)	Sortal particulars construed as manifesting in collected mode
Talk of divided modes	Specification of degree of dividedness in collected mode
Two operational types of objective valuation	Unitary element assigned by operational means

Ẹyọ (ohun)	*Ẹyọ (ìdì)*
Units made analogous to integers of number	Units made analogous to integers of number
Kà	*Wọ̀n*

derlies counting and measuring. However, if one is happy to acknowledge
that the basis of translation is physical operation, and not be tempted to
read more into it than merely that, we can see that these translations are
apt.

I have postulated the existence of perceptual universals, the percepts
that people seem to allude to when valuing as part of the exchange of
goods. These percepts are named by quality nouns in English: weight,
volume, length, and so forth. How do these percepts come into the conceptual scheme of Yoruba language valuation? They enter the process operationally through selection of the type of unit. Use of *ẹyọ (ohun)* results
in a process that often looks like counting, and in this case we can recognize a parallel with the use of the quality of numerosity in the counting
process of English language quantification. In the different types of *ẹyọ
(ìdì)*, parallels with the use of units of the various qualities that lie behind
English language measuring can be recognized. *Ẹyọ (ìdì)* may be of many
operational types. They may be types of calabashes when palm wine is
being quantified. Arranged in increasing size, the names of different *ẹyọ
(ìdì)* for palm wine are *igó, ahá, àdému, garawa.*[15] (In English language
terms, it would be said that these measure volume.) One may *kà* cloth if
one is buying or selling set lengths such as are produced by hand weavers.
In another context (buying a specified length from a bolt), one may *wọ̀n*
cloth. One may use the following *ẹyọ (ìdì)*, depending on one's needs:
kànkó (literally "knock"), the length of the upper surface of the middle
finger laid along a surface from fingertip to second knuckle. This unit is
often used by *fìla* (cap) makers. Still concerned with what an English-

speaking person would call measures of length, there is *gbaga,* the span from thumbtip to middle fingertip, and *ìgbònwó,* the length from wrist to elbow. To *wòn* land the *ẹyọ (ìdì)* of *okùn* (string) is used; a boundary may be created by the use of string. More usually, land is *wòn* using *ẹyọ (ìdì)* of one yam mound (a large mound of soil in which yam tubers are planted), which is characterized as *poro kan* (one sideways leg span) by *ẹsẹ kan* (the length of a step).

Number Talk

Number comes from our talk of our material practices. We talk of things as making up the world, and we say these things have features. In English, the things we talk about have qualities of various sorts to varying degrees, and in Yoruba, the things we talk about have mode—they manifest in the here/now or then/there in infinitely varied ways. It is through our talk of the features that we say are manifested by the things we say there are, that we come to number.

In English language talk, number is a concept that organizes the concept of qualities, which in turn organizes the concept of spatiotemporal particulars. The abstracting sequence of generalizing through which English speakers come to number is summarized in table 9.3.

An English-speaking person might use natural number in quantification and say, "Four oranges are there in the bowl," or "I would like one

Table 9.3. Sequence of Abstraction in Coming to Number in English
Language Quantification

Things here/now and then/there
Spatiotemporal sections of matter
│
Abstraction of spatiotemporal sections of matter
│
Qualities
Attribution of certain extents of characteristics of various types
│
Abstraction of extending qualities
│
Numbers
Attribution of an extent along the infinite continuum of plurality of units of a quality

kilogram of oranges, please," and she would feel confident that she is talking about what there is in the world, in a precisely correct way. In the first case, she would be using units of the quality numerosity, and in the second, units of the quality of mass. Numbers introduce precision, and she and other English speakers know whether English language talk using natural numbers is right or not. When you speak English, there is a right and a wrong way to talk of things and their qualities, and to use numerals in that talk. There is a right and a wrong way, not because the world is organized in a particular way, but because English-speaking people engage in particular organizing practices and *say* it is organized in that way.

Yoruba speakers talk of a particular sort of continuous matter, an object spread infinitely over the here/now–then/there. In generalizing this category, they come to talk of mode, a "slice" of a sortal object, bounded by the here/now or then/there. In organizing these slices of the here/now–then/there continuum, Yoruba speakers come to number. Number, in Yoruba language talk, is a degree of dividedness. Yoruba language number reports degrees of dividedness within here/now–then/there slices of sortal particulars. In Yoruba language talk, number organizes mode, which in turn organizes sortal particulars. I have summarized this series of abstracting generalizations in table 9.4.

A Yoruba fruit seller will quote prices for oranges in groups of five *(ọsọ̀n márùún)* or twenty *(ogún ọsọ̀n)*, for in Yorubaland that is how oranges are generally sold. The orange seller is no doubt confident that in talking of *ọsọ̀n márùún,* which translates literally as "orangematter in the

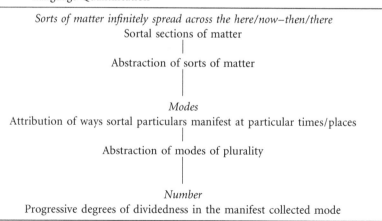

Table 9.4. Sequence of Abstraction in Coming to Number in Yoruba
Language Quantification

Sorts of matter infinitely spread across the here/now–then/there
Sortal sections of matter
|
Abstraction of sorts of matter

Modes
Attribution of ways sortal particulars manifest at particular times/places
|
Abstraction of modes of plurality

Number
Progressive degrees of dividedness in the manifest collected mode

mode collected in the mode five," she is talking of the world as it really is; she is quite confident of the correctness of her way of valuing oranges. If you are a Yoruba speaker, you understand something quite precise if you understand the sentence "Ọsọ̀n márùún," and you must be correct when you speak of these things, or else what you say is meaningless.

The Recursions That Constitute English Language Numerals and Yoruba Language Numerals

I have spent some time talking of how English speakers come to have spatiotemporal particulars with various qualities in their world and how Yoruba speakers have sortal particulars in various modes in theirs, but so far I have glossed over the second level of abstraction, which is wrought in coming to "natural" number—from modes or qualities, as the case may be, to numbers. Let me now consider what it is to abstract modes in coming to natural number in using Yoruba numerals and to abstract qualities in coming to natural number in using English numerals.

Number originates in the practice of material tallying—modeling an event or episode in the world with a material encoding process. We can see this as using a finger to encode the passing of a sheep through a gate, the placing of a pebble to encode the pointing at a soldier, or the engraving of a line on a bone or a piece of wood to record the filling of a vessel with grain.

Encoding these practices of material ordering in a sequence of words is to develop a numeral system. A numeral system encodes the patterns made in the process of material ordering. It derives, not from encoding regularities of the material world, but from encoding social practices in organizing the material world. Since fingers and toes are the handiest model, we find that the patterns that have been made in encoding material practices derive from our digits.

A numeral system is characterized by having a sequential base pattern and recursivity. Numerals constitute an infinite series by having a base about which repetition occurs, and a set of rules by which new elements are generated. They encode material practices that are themselves the product of the working together of several semiotic systems: a signing system of language, a system of body signs, and a system of material organization. Fingers and toes are used to encode events resulting in unitized matter, and in talk, language is used to encode signs created in both these signing systems. I illustrate this point through comparing the characteristics of the numeral system associated with English language quantification and the numeral system associated with Yoruba language discourse of *iye*.

The contemporary numeral system, which has developed in association with Indo-European languages like English, has ten as its base; in other words, ten is the point in the series that marks the end of the basic set of numerals. As each ten is reached, the basic series starts again, each time recording in the numeral how many tens have been passed. The rule by which new elements are devised is addition of single units and base-ten units.

The Yoruba language numeration system is a multibase numeral sequence that makes constant backward and forward reference to multiples of the bases. The most important base of the system is *ogún* (twenty); it is a vigesimal (base twenty) system. *Àrùún* (five) and *èwá* (ten) provide points at which the twenties are broken up, so we can say that the system is secondarily quinary and decimal. The rules for recursion in the Yoruba numeral system make little use of addition; the processes of multiplication and subtraction are more important. Only the smaller numbers have obligatory forms; the form of higher numerals depends on the process by which the numeral is devised.

One can easily imagine the involvement of fingers and toes in tally keeping in a nonlinguistic way. One separated finger codes for one separated item. But if we then extend the coding operation and say a word that codes for the finger or toe, we have done something much more complex, and ended up with a code that is much more useful than the material code of fingers and toes.[16] In saying a word as a finger is held up to code for an item involved in some event, we understand that the word we say does not name either the item or the finger. It names a position in a progression. Numerals are words that code for a position in a series. Yet even to say that numerals are a linguistic code with which one may record how far through the series of fingers and toes we have progressed, and how many times we have done it, is to make a seductively simple description of a complex coding affair. Let us examine this apparently simple step more closely and carefully trace the links between the material world and the numeral code.

The fact that a tally is being created already implies two things: the matter under consideration is taken to be of "a sort," and the stuff is separated into individuated sections (items or units) so that it can be the focus of an event within a more complex social occasion, an event that can be encoded by holding up a finger.

Systems of Recursion, Systems of Unitizing, and Systems of Designating Work Together

In comparing Yoruba and English language numeral systems, we see that a numeral-generation system that has addition of units of ones and units of tens as the generative rule is associated with talking of material objects as spatiotemporal particulars. In a form of life where sortal particulars are talked of, the numeral system has a base of twenty within which subsidiary bases of ten and five enable progression by ones through the processes of multiplication, subtraction, and addition. I suggest that this association is significant. The types of material objects talked about will determine the recursive rules of a numeration system through determining the way fingers (and toes), taken as a material code, may be spoken of. For when you come to name the finger that you have reached in the scale of fingers to code for position in a progression, the types of things that a language postulates as being in the world determines the way that an individuated separate finger may be designated. In a numeral system, the practice of referring and the practice of tallying are rolled up together.

When the primary objects talked of are spatiotemporal objects, the linguistic code explicitly differentiates fingers from one another. When the primary objects talked of are sortal particulars, a linguistic code to report position on the finger-toe scale must necessarily be more complex. The entire scale (twenty fingers and toes), the digital complement of a person (a sortal particular), must be the starting point. In dividing and subdividing the digital complement of a person, one starts with a sortal particular (a person) and comes to identify a specific individuated finger or toe, a position on the finger-toe scale.

It is my contention that that the multiplicative/subtractive/additive Yoruba numeral-generation system is a consequence of talking of sortal particulars, just as the additive numeral-generation system of English is the consequence of talking of spatiotemporal particulars. When fingers are held up in code for separated material items, their separateness and spatiotemporal uniqueness is inevitably commented upon by English speakers in designating. It is a relatively simple matter in English to report the addition of more individuated sections. In the English language numeral-generation system, one can infer an activity of moving along the set of fingers and, on reaching the end, designating the collection as an individuated section, a collection of ten.

With the primary categorical distinctions implicit in using Yoruba language, the fact that finger/toe matter naturally manifests as collections of twenty individuated items and is naturally subdivided into collections of ten and five is relevant. A sortal particular, a person, coincides with the manifestation of finger/toe-matter in this way. One specific unit of finger/toe matter can be specified *ogún*, and it then becomes reasonably straightforward to specify a certain finger or toe by using the natural divisions of a set of fingers and toes.

The Yoruba language numeral-generation rules can be taken as modeling the following sequence of specifications. Beginning with the finger/toe complement of a person gives the major *ogún* (twenty). Shifting from one vigesimal to the next codes for "starting a new set of fingers and toes"—literally "placing out a new set of twenty" in Yoruba, for example sixty (20 × 3), *ọgọ́ta*, is twenty placed three times. Specifying "l'ẹẹ́wa ó dín" (reduce by ten) signifies "hands only," and the additional specification "l'árùún ó dín" (reduce by five) signifies "one hand only." At this point, we have ($-5 - 10 + (20 \times 3)$), *márùúndínláàádọ́ta*. Continuing in this vein, the next specification is "reduce the reduction by ones" ($-4 - 10 + (20 \times 3)$), *mẹ̀rìndínláàádọ́ta*, and along the fingers and so on to the next hand and along its fingers ($+1 - 10 + (20 \times 3)$), *mọ́kọ̀nlél-áàádọ́ta*. Next specify "toes (one foot)" ($-5 + (20 \times 3)$), *márùúndínlọ́-gọ́ta;* then "reduce the reduction by ones, working along the toes" to "add the toes of the other foot by ones" ($+4 + (20 \times 3)$), *mẹ̀rìnlélọ́gọ́ta*. Notice that the first *ogún* is not specified before starting; one works toward the first *ogún*, not back from it. Presumably, the numerator starts with his or her own fingers/toes. Yoruba numeral generation seems to imply a picture of numerals nested within each other: digits nested in persons, fingers and toes in hands and feet.

This is very different from the picture we get with the base-ten numerals used in modern quantifying and mathematics in English. There, with twenty implying "the fingers of two hands gone through twice," twenty-one begins again: thumb of left hand held up; twenty-two, forefinger of left hand; twenty-three, tall man of left hand; and so on to twenty-six, thumb of right hand; and on up to thirty, which is three sets of two hands. What we see in English is a linear passing along the fingers until the end, keeping tally, and doing it again. Each integer is equivalent and related to the one before and the one after it in a linear array. Demonstrably, both modern base-ten numeration in English and Yoruba numeration are tallying recursions.

Interestingly, we find that these contrasting patterns that we see generated in English and Yoruba language numerations echo a contrast we find in the accounts of number given by Western mathematicians. Von Neumann and Zermelo seek to explicate number by identifying the referents of number words, and they have come up with quite different explications. Their enterprise is very different from the one that engages me;[17] nevertheless their differing accounts of number can be useful in helping us see the contrast between English language and Yoruba language numeral recursions more clearly.

As an English language speaker, von Neumann's account seems to me to be intuitively correct. Cardinality on his account seems to encode a one-by-one collection of predecessors. Von Neumann has the members of an n-membered set paired with the first n numbers of the series of numbers. For von Neumann, zero is $\varnothing$, the empty set. The set that contains the empty set as its sole member is $\{\varnothing\}$, one; the successor of this number is $\{\varnothing, \{\varnothing\}\}$, or two; and three is the set of all sets smaller than three, $\{\varnothing, \{\varnothing\}, \{\varnothing, \{\varnothing\}\}\}$. The successor of any number in von Neumann's version is generated by adding the successor of zero, that is, one. A number in this version is the last number of the series reached through one-by-one progression. A numeral names the point at which the progression ceases.

In contrast, I suggest that a Yoruba speaker would choose Zermelo's account of number as correct. Here the number n is a single-membered set; the single member of the set is $n - 1$. Zermelo has zero as $\varnothing$, a set with no members. Then 1 has the empty set as its sole member, $\{\varnothing\}$. The set that contains the unit set, 1, as its sole member is 2, $\{\{\varnothing\}\}$. Three is the set that has two as its sole member, $\{\{\{\varnothing\}\}\}$. In this version, each number is totally subsumed by its successor, and any one number has a unified nature. I contend that, for a Yoruba speaker, the model of number that would jump out as the intuitively correct account would be this one, for Yoruba language numbers carry the flavor of a multiply divided whole; there is no sense of a linear stretching toward the infinite here.

The two models agree in overall structure in identifying number; each model is demonstrably a recursive progression, although importantly for logicists, for von Neumann (and I suggest English language number) the set 14 has fourteen members, while for Zermelo (and I suggest Yoruba language number) 14 has one member only. While this is a significant difference for logicists, for ordinary users of number, with no interest in models of cardinality, the difference is irrelevant.

This interpretation of the Yoruba language numeral code describes a way of proceeding so that positions in a progression may be reported in a way that is not tied to stimulus situations, that is, use of people's actual fingers and toes, while speaking a language that does not utilize spatiotemporal position in defining material objects to talk of.

I have related differences in the common designating categories of English language and Yoruba language to the readily apparent differences between the traditional Yoruba numeral sequence and the contemporary Western numeral sequence. It should not be inferred from this, however, that a link is necessary between designating category type and the base and generative rules of a numeral sequence used in that language. The way the world is rendered in talk may determine the type of numeral system conjured into existence in that talk, but once in existence, the numeral system takes on a life of its own. In turn, it contributes to the remaking of the form of life in which it has existence and might be remade itself through the contribution it makes.

Conclusion

I have unpacked natural number as used in the English language and in the Yoruba language and shown them as systems of abstraction created in two disparate systems of categorization. I have argued that the basis of the difference between the two chains of abstraction is the difference between Yoruba and English in the types of actions that bodies engage in that is coded in coming to predicate (use verbs) and so to designate in the language. The sort of construct that natural number is in a particular language has been determined by prior (and ancient) decisions about coding interaction in the material world. They are tied up with what are the now taken-for-granted rules of using verbs and their subjects in that language. English language natural number and Yoruba language natural number are constituted by both linguistic and nonlinguistic concepts. They are a chain of abstracting practices: practices of referring, practices of defining units of matter, and practices of tallying units. They both provide a means by which value is ascribed to the stuff of the world. It is the ability to function as a tool in the context of certain types of work that marks natural number, and in this the Yoruba and English language systems of abstraction perform equally well. Irrespective of how they encode plurality, they work.

Yet as constructs they differ, resulting in separate worlds of symbolizing logic.

Number is a tool made within a form of life that in turn contributes to the remaking of that form of life. Number encoding one version of cardinality will contribute in different ways and achieve different things than number encoding another. At one point in our history, natural number was the subject of laboring by people to produce meaning. We have forgotten all that as we go on using it as though it were a fact of the material world, rather than of the social world.

In making this case study of natural number, I have been attempting to relocate and recontextualize it in labor and forgetfulness. Those who in the past labored to make new meanings inevitably changed the system. Once meanings are made, they are viewed as natural and the processes by which they were conjured into existence are taken for granted; their status as products of human labor is forgotten. They have become black boxes. It is at that point that the product becomes the grammar of the system.

Natural number is a part of grammar. It has been naturalized. That is why most people fail to see it as a social product. In the forms of life in which it has existence, number has taken on a life of its own and is capable of subordinating the labor of those who work with it. People are driven, as it were, by number in their use of it. I suggest that it is in this way that mathematics derives its seeming inexorability and inevitability. Natural number is a historical product, and it is we who have both made it and rendered its making invisible.

Chapter Ten

Decomposing Predicating-Designating as Representing

*I*n two famous and controversial papers that are still being referred to some thirty years after their publication,[1] Robin Horton sought a definitive elaboration of the similarity and the difference between "African traditional thought and Western science." For Horton, the crux of their similarity was that both "systems of thought" worked through theoretical elaboration—systems of symbolic representations that referred to the material world. Horton has both science and "African traditional thought" as universalist, where certainty is achieved in a degree of correspondence between physical reality and these theoretical systems. The essential difference, as Horton saw it, was that "African traditional thought" was committed to a "closed" certainty, which explains its backwardness, while "Western science" struggled with the dilemmas associated with being "open." "Openness" involved some uncertainty, but allowed for the possibility of progress.

At the center of Horton's pivotal contrast between "closed" and "open" is the working of language. In a "closed" system, words and things are forever stuck fast together, routinely mistaken for one another. The system of representation, understood as a form of fetishism,[2] is unchanging and unchangeable. Knowledge and those who authorize it are totally "closed" to criticism. In an "open" system, by contrast, words have become detached from things, so that changes in theories become possible.

Horton sees the possibility of changes in systems of representation as the defining characteristic of "open" science. Such changes constitute the progress that science is supposed to make in its development of theories. Systems of representation are increasingly accurate pictures of the material world. Scientists and their accounts are "open" to systematic critique.

Underlying Horton's comparing and contrasting is the assumption that knowledge and knowing necessitate a given world and, quite separately, its representation. Being symmetrical, and departing from anthropological orthodoxy, he attributed foundationism, with its specific ontology, equally to African traditional thought and Western science. In attributing to African traditions of knowing a view that has a material world and knowledge of it as a priori separate, Horton offered no warrant, although the notion is highly contentious. Despite its being clearly an empirical matter, he presented no evidence for this assumption.

Horton attributed a primitive form of foundationism to African traditional thought; it had neither philosophy nor logic.

[T]here is a sense in which [African] thought includes among its accomplishments neither Logic nor Philosophy. . . . traditional thought has tended to get on with the work of explanation without pausing for reflection upon the nature of rules for this work. . . . The traditional thinker, because he is unable to imagine alternatives to his established theories and classifications can never start to formulate generalized norms of reasoning and knowing. . . . [In contrast] the "open" predicament [of science] makes it inevitable they must appear. For when the thinker can see the possibility of alternatives to his established idea-system, the question of choice at once arises and the development of norms governing such choice cannot be far behind.[3]

In part, it is infamy generated in the outrage of the African academy at this assessment, rather than fame, that makes Horton's papers well known. Probably more significant in generating the attention these papers received was the way in which they seemed to intervene in a crisis in the philosophy of science that was raging at the time of their publication. Indeed, it seems that this controversy had in part inspired the papers.

The crisis in the philosophy of science grew out of controversy over how theory change in science should be understood. The orthodox view, which Horton seemed to endorse, was that scientific theories changed through rational argument by scientists. These arguments exemplified scientists' commitment to achieving certain representations (theories) though their "higher level" commitment to uncertainty. Controversy was sparked when, with the publication of Thomas Kuhn's *Structure of Scientific Revolutions* in 1962, this comfortable assumption about the working of science was no longer tenable. Invoking particular episodes in the history of science, Kuhn portrayed scientists as lurching from old certainty to new certainty through short-lived periods of uncertainty best understood as irrational revolutions. On both sides of this controversy, certainty was seen as discursively generated; that is, knowledge authorities accept as legitimate particular representations and not others. The assumption here is that in "Western science" the connections the authorities legitimate between words and things are "truer" than those of, say, "traditional African thought," because science, being committed to "openness," has the possibility of progressing to better truths. Kuhn agreed with the orthodox position that in science the authorities periodically come to authorize new connections as certain representations, but against orthodoxy, he insisted that there was nothing particularly scientific about the way they accomplished this change.

This controversy over theory change, in which Horton sought to have a voice, intersected in complex ways with another ongoing disagreement within the philosophy of science. This second debate was between those who saw truth as located in theories authorized because of their correspondence to the world, and those who had certainty located in the coherences of systems of representation variously originating in founding practices that generated differing forms of representation. On this division, Thomas Kuhn's supporters and his opponents are universalists who have theories referring to natural, given categories, and opposing both these groups are relativists who refuse given categories, having them instead as the outcome of a social practice.

Some twenty years after Horton made his intervention in the philosophy of science by mobilizing arguments from African contexts, I attempted a similar move in this related though distinct controversy. To some degree, I distanced myself from Horton, yet I also unthinkingly followed him in my assumption that there must be foundations. My view assumed as general the existence of a foundation for knowing and knowledge, but opposed the assumption that certainty lay in a single authorized account of the structure of material world contained in (replaceable) scientific theories. On the contrary, as I saw it, certainty lay in coherence constituted within the symbolic representations of general categories created in past social practice. Locating certainty there seems to offer the optimistic possibility of discursively elaborating differences in logics.

That, I felt, would enable me to understand the workings of "systems of thought," and by implication how they might work together in struggling for better futures. Having laid out this theory in the paper of chapter 9, all that remained was to wrap it up. That, as you already know, never happened.

Seeing My Relativist Literalizing

I now understand the arguments of that paper as flawed by their dependence on self-reference. The paper that constitutes chapter 9 makes three categorical errors of literalization that secure the differences I show. I have come to recognize my relativist theory as a comedy of errors. The literalizations have the effect, among other things, of explaining away exactly what my project sought to explain—the realities of managing the evidently complex arena of numbering in Yoruba classrooms.

Paradoxically, the arguments depend on a form of fetishizing of exactly the type that Horton attributed to "traditional African thought." In this chapter, I reveal in detail the contradictory moments making up the literalizing that secures the argument about difference in predicating-designating categories in the first section of the paper.[4] This is significant because it concerns the making of a boundary that is the fundament of a foundationist imaginary—the separation of material world and its representation. The boundary between the predicating and designating categories, and what it is that they represent, is foundationism's sacred boundary between things and words. Transgression of the boundary in the form of a literalizing category mistake is proscribed in foundationism. This, Horton argued, is what "primitives" do. This is the mistake, he implies, that is responsible for all Africa's miseries.

I show that my argument about difference between English and Yoruba predicating-designating categories is secured by literalization of the figure matter/space/time. This literalization is not restricted to my relativist argument. As an extension of Quine's universalist formulation, my literalization remakes Quine's within a relativist frame. What our elaborate argument does is secure a boundary that forbids and proscribes literalization *with* a literalization. A boundary that may not be transgressed in foundationism is rendered through its transgression.

I am not the first in science studies to point in this direction and draw attention to the elements of illusion embedded in foundationism. In his polemic essay *We Have Never Been Modern,* Bruno Latour details what he calls the opposed "purifying" and "translating" elements that are equally essential for science. In the purifying, the foundation is established; in the translating, particular instantiations are "discovered." He too shows foundationist claims to certainty as incredible; look closely and it is revealed as a well-disguised hall of mirrors.

All three elements of the tightly linked chains of abstracting logic that I show in chapter 9 dissolve. Foundationist claims that representations may be evaluated as certain/uncertain dissolve. What showing the literalizing does is show that there is nothing to be certain or uncertain about. Foundationism's claims to explain all possible worlds must be refused. Nevertheless, recognition of foundationist, scientific explanation's dependence on self-reference does not destroy a much more modest claim. In the sorts of ordered/ordering microworlds we see in science and pervasively in modern times and places, foundationism's claims to offer valid explanations will hold. In much the same way, self-reference does not invalidate the workings of a program in a computer. If science's explanations are to be made as general claims, its microworlds must be infinitely extended.[5] The implication is that worlds must be remade before they can be explained. I have suggested that a different strategy for connecting up times and places is through telling stories of the lives of objects and how these lives proceed in reliably managed worlds.

The Common Human Practice of Predicating-Designating Generates the Foundation

Clearly signaled in chapter 9's first section is the constitution of the general and particular domains. This separating—constituting the two domains—is accomplished in the first nine paragraphs conjuring up the general form of the common human practice of predicating-designating. My account of the general form of predicating-designating was developed from Quine. It is taken to result from the linking of two quite ordinary childish classifying activities. The classifying activity distinguishes perceived similarity and perceived difference, and extends this to the heralding of these two classifications with utterances—words.

Because they express a fundamental distinction—between similarity and difference—words coding the two categories can be linked. In their joining, a whole new level of category appears. What constitutes these primal categories of similarity and difference are the givens of matter, space, and time, and in the joining of the two categories matter and space-time are set against each other in this way or in that to make the foundation predicating categories and the foundation designating categories.

Like Quine, I took matter, space, and time as found, the given things in the world. My analysis went beyond Quine's to assert the possibility of a general, nonspecific predicating-designating. This was based on my contention that either of two forms of coding and joining coding items through juxtaposition are possible.

In one way of joining, similarity is coded as action and difference as object. Similarity coding action implies that action is the change between "manifesting in space-time characteristically as a particular form of matter," and "*not* manifesting in space-time characteristically as a particular form of matter." Action here is the going from manifesting to not manifesting, or vice versa. Difference coding object focuses on the inherent differences between the types of matter that might engage in such an action. For example, "dogmatter manifesting characteristically" is how similarity becomes action; the difference between dogmatter and catmatter is what effects the object dogmatter. This is the general form I attributed to Yoruba predicating-designating.

In the other way of achieving a similar result, similarity codes object and difference codes action. Here the similarity of matter is what connects objects as objects. Difference is the same object manifesting in different space-time locations. For example, "dogs" are space-time objects, and dog appearing inside the door now and dog appearing outside the door later, that is, in different space-time locations, is difference. This is how I saw the contrast rendered in English talk.

I might have said that Quine, not having had the benefit of a prolonged acquaintance with a radically different language, as he calls languages such as Yoruba, has a limited imagination here. Understandably, he failed to see that the former way of arranging the terms is feasible. In turn, he failed to realize that the account he gives of predicating-designating can be reconstrued to account a generic rendering of the given categories of the material world as object and action, in achieving predicating and designating.

The foundation-setting work reveals a natural world in which the possibility of objects (of an unspecified sort), which can engage in possible actions (of an unspecified sort), has been set up. This "nonspecifically socialized nature" becomes the domain of the general. The symbolizing domain is also set, started by the generation of a particular sort of predicator and a particular sort of designating category—the nature of which was determined in the particular disposition of the two different types of heralding terms in their joining.

Finding the Instantiations: First, Getting beyond Indeterminacy in Translation

Now, to show the opposing work, I turn to considering how the partic-ular versions of "predicator" and "designating category" are generated in my text as *found* particulars. As I puzzled about how Yoruba-speaking children could quantify quite outside a logic that evoked qualities, I began to consider the idea that talking in English somehow created a climate in which qualities were necessary for quantifying, whereas talking in Yoruba evoked the climate for alternative concepts, ones that fitted with Yoruba numerals functioning as mode nouns. I was inspired W. V. O. Quine here.[6] Yet, while Quine inspired me, his argument on indeterminacy in translation denied any possibility of a useful comparative investigation of Yoruba and English logics as we see them expressed in quantification, and hence nipped in the bud any idea of a general theory of many relative logics.

For Quine, "language is a social art"[7]—an enterprise of manipulating symbols—and how the symbols map onto the material world is inscruta-ble, unknowable. For Quine, the world presents universally as stimulus— he is a universalist on my way of reading foundationists. The closest con-nection between the symbolic world of language and the material world, which causes sensations, lies between "occasion sentences" and "stimulus meanings." While he recognizes that we might see a correlation between these, significantly, that is all it is. There is no connection across the gap between the symbolic world of languages and the material world of which we are sensible. The gap remains unbridged and unbridgeable. It is this that ensures indeterminacy in translation: since we cannot know how the categories of either English or Yoruba map onto the material world, we cannot know how they map onto each other. According to Quine, I had gone as far as I could go.

In making this argument, Quine gives us a picture of a linguist trying to develop a translation manual for two languages that have never before been translated, and that are not connected by the chain of interpreters that in practice connect most languages. Quine muses over the way such a linguist might work.[8] We can imagine them sitting companionably on a hillside in a gathering twilight. The native-speaking informant mutters "Gavagai" when a rabbit runs by. The linguist, as a foreigner, develops the hypothesis that what in English is called a rabbit is called "gavagai" in his informant's language. But, Quine points out, such a judgment would be hasty. Even supposing, says Quine, that the native experiences and notes an area of dark that moves across a stable, lighter background in his field of vision (a stimulus to which the linguist responds with the utterance "Lo, a rabbit"), many objects other than a whole and enduring rabbit are compatible with the stimulus.

> Stimulus synonymy of the occasion sentences "Gavagai" and "Rabbit" does not even guarantee that "gavagai" and "rabbit" are coextensive terms.
>
> For, consider "gavagai," who knows but what the objects to which this term applies are not rabbits after all but brief temporal segments of rabbits? In either event the stimulus situations which prompt assent to "Gavagai" would be the same as for "Rabbit." Or perhaps the objects to which "gavagai" applies are all and sundry undetached parts of rabbits: again the stimulus meanings would register no difference. When, from the sameness of stimulus meanings of "Gavagai" and "Rabbit" the linguist leaps to the conclusion that a "gavagai" is a whole and enduring rabbit, he is just taking for granted that the native is enough like us to have a brief general term for rabbit and no brief general term for rabbit stages or parts.
>
> A further alternative likewise compatible with the same old stimulus meaning is to take "gavagai" as a singular term naming the fusion . . . of all rabbits. Thus even the distinction between general and singular term is independent of stimulus meaning. . . . And a still further alternative in the case of "gavagai" is to take it as a singular term naming a recurring universal, rabbithood. The distinction between concrete and abstract object, as well as between general and singular term, is independent of stimulus meaning.[9]

These various types of categories that Quine identifies as being consistent with the English use of the term "rabbit" in "Look, a rabbit running by" derive from using boundaries, which Quine takes as found, for categorizing the world: matter with various sets of characteristics, set in space and time. Quine identifies five different possibilities, in the quotation above. The informant might, says Quine, be evoking a category by binary classification over rabbitness/nonrabbitness—which is in orthodox terms an abstract object engaging the notion of the quality of rabbitness. He might, alternatively, be evoking all that matter, infinitely scattered through space and time, that exhibits the characteristics of rabbitness— what we might call rabbitmatter. It may also be, says Quine, that the informant is evoking rabbitmatter situated with respect to time, or rabbitmatter situated with respect to space. It is possible, too, he says, that the informant is evoking rabbitmatter that extends through space, endures through time, and is enclosed by a boundary. In this case only would he be evoking the same type of category as the English-speaking linguist, when he says, "A rabbit ran by."

I have listed the five types of categories Quine identifies as possible candidates in table 10.1, also describing his categories in terms that mean more to me.

It is important to distinguish between the "rabbit" of the English sentence "Rabbit!" and the "rabbit" of the English sentence "Look, a rabbit is running by." The first use might be called a heralding or an announcing; no particular category is involved here. But in this way of understanding language as a symbolic code, the second use designates, and designating evokes a particular category as the outcome of the activity of predicating.

According to Quine, there is no way the designating categories of two languages can be mapped determinately, because the only way to do that would be to proceed through the way each symbolic system (i.e., language) maps onto the material world, a mapping that is unknowable be-

Table 10.1. Quine's Categories of Particulars That Might Be Evoked in Predication

Type of Particular (Quine's Terms)	Type of Particular (My Terms)	Implied Category
1. "Rabbit stages, brief temporal segments of rabbits"	Time segment qualified as rabbitty: a temporal particular	Matter with a particular set of characteristics situated with respect to time
2. "All and sundry undetached parts of rabbits"	A spatial extension qualified as rabbitty: a spatial particular	Matter with a particular set of characteristics situated with respect to space
3. "A whole and enduring rabbit"	A spatiotemporal extension qualified as rabbitty: a spatiotemporal particular	Matter with a particular set of characteristics situated with respect to space and time
4. "A fusion [of] that single though discontinuous portion of the spatiotemporal world that consists of rabbits"	Rabbitmatter: a sortal particular	Matter with a particular set of characteristics
5. "Rabbithood"	Rabbitness: an abstract qualitative concept	A set of characteristics

Source: W. V. O. Quine, *Word and Object* (Cambridge: MIT Press, 1960), 51.

cause scrutiny of the categorizations inherent in the material world is impossible. Quine's purpose is not to deny that translation is feasible in a practical sense. His point is that translation can never be determinate. For him inevitably, translation is the imposition, to a greater or lesser extent, of the parochial linguistic habits of the speakers of one language onto another language. And, since the extent of this imposition can never be known, Quine suggests that it is best for the linguist to assume that the foreigner is like himself, and opt for the banal, rather than the exotic, translation.[10]

This is the second point where I refuse some of Quine's limitations, disagreeing with him over indeterminacy in translation. While agreeing with him on the issue of the inscrutability of categories generated in predicating-designating, I insist on effability, justifying this in chapter 9, note 3. Effability is not an issue for Quine because he has entirely missed that difference and similarity might be coded for action and object in more ways than one.

I adopted the account of predicating-designating developed from Quine as a *general* account of predicating-designating. Unlike Quine, I took it that the juxtaposition of the two types of heralding terms can be achieved in more than one way, evoking objects and actions of more than one type. Missing this, Quine focuses on the impossibility of the process of scrutiny and assumes that determinate translation could proceed only in that way. I agree with Quine that the categories generated in any particular form of predicating-designating are inscrutable. Contra Quine, pointing to comparative effability of the categories rendered in predicating-designating, I insist that the categorizing of one language can be talked about both in it and in other languages. Since the categorizing arises in the variable workings of languages as symbolizing processes, there is no pretense of reflecting the workings of the material world.

Finding the Instantiations: Second, Doing a Linguistic Experiment

Satisfied on the possibility of comparatively laying out the categories generated in predicating-designating in English and Yoruba, I showed myself setting about examining the workings of verbs and subjects in sentences. Agreeing with Quine over the range of possible designating categories (see table 10.1), I found it to be a relatively simple matter to identify, from examining the workings of verbs and their subjects, the type of category generated in English language and Yoruba language predicating-designating. I identified number 3 in the list—spatiotemporal particulars as generated in English—and number 4—sortal particulars as generated in Yoruba.

Using verbs and their subjects and objects in invented English sentences and invented Yoruba sentences as translators, I managed the alternative conflations of designating category and sortal particular in Yoruba, and between designating category and spatiotemporal particular in English. A sortal particular is matter of a particular kind spread across all space-time frames, but for which, in a modification, a frame here/now or then/there can be specified. A spatiotemporal particular is matter that is here/now or then/there, of which a set of given characteristics may be specified in a qualification.

Translating and paraphrasing were central. In the invented English sentences "It is a hoe"/"It is water" and the Yoruba sentences "Ọkọ́ ni ó jẹ́"/"Omi ni ó jẹ́," paraphrasing words working in the verb position in sentences—*is*/*jẹ́*—to other words, I showed how the foregrounding/backgrounding in the working verbs condensed a specific presentation as nouns in the subject position in sentences, using as a translator the figure of matter set in empty space-time.

> "It is a hoe" may be taken as meaning "The thing here and now is a hoe." Similarly, in saying "It is water," a person could be taken as meaning "The thing here and now is (a glass of) water." "It" evokes a here-and-nowness. We may paraphrase the designating term "it" as "this spatiotemporal particular." The speaker is identifying a section of the material world that is here-and-now. Its here-and-nowness is what defines it as a particular. "It" is a separated section of matter that extends in space and endures through time.

Also,

> *Ọkọ́* and *omi* name what the sentences "Ọkọ́ ni ó jẹ́" and "Omi ni ó jẹ́" are about; *ọkọ́* and *omi* are designating terms. But what is designated? What is the category? The meaning we take from the verb will help in identifying this. So too will the introducer, which points to the verb phrase as emphasizer. *Ó jẹ́* is generally translated as "it is"; however, the verb *jẹ́* has a different meaning than *is* in English. Rather, *jẹ́* implies "exists or manifests with its inherent characteristics."

Through the paraphrasing, I make an elision between nouns that are subjects of English sentences, and the picture of spatiotemporal things, and nouns in the subject position in Yoruba sentences, and the image of sortal things. In the translations, I relied on the fact that any competent English speaker or competent bilingual Yoruba/English speaker could adjudicate the paraphrases, elisions, and translations I made between English and Yoruba. All these elaborate elisions, translations, and paraphrases become the "method" of my searching, and I consider that I have "found" two particular sorts of predicating-designating categories.

Juxtaposing the "Found" Categories and the Foundation Categories

When I contrast the "found" particulars, the various sorts of categories my experiment "discovered," with the general category, we see that something bizarre has happened. The terms used to define the general category—matter, space, and time—are also used to describe the particular categories. "Found" categories and foundation categories are one and the same, but have been contrived as categorically separate. The sameness and difference contrast, through which the general possibility of predicating-designating is generated, is set through the notion of matter in a space-time frame, just as the particulars are set through these same notions. Things *are* their representations.

Along with Quine (and many, many others), taking the image of matter/space/time literally, I secure the notion that predicating-designating is the first step in representing the world, and in doing that I confidently render accounts of English and Yoruba for all English and Yoruba speakers. Like English, Yoruba represents a world out there. A metaphysics is legislated, giving words a special and nonmaterial status. Words are made as a priori nonmaterial, and materiality is defined as nonlinguistic. I have used a style of reasoning set within a specific metaphysics. It should come as no surprise that the outcome of using that metaphysics is its remaking. It is a style of reasoning that depends on literalizing to secure its arguments and, demonstrating that literalizing, dissolves that metaphysics' claims over the possibility of certainty in its explanations.

Through literalizing—having the figure of matter/space/time as at once a thing (or all things) and its representation, my foundationist account of predicating-designating rules that words cannot be things, and things cannot be words—despite the fact that what you have just read and are still reading is clearly both things and words simultaneously. Still that is a commonplace paradox; many take it for granted. My account of the little linguistic experiment is a way of dealing with complexity: the funny and paradoxical ways of collective going on in contemporary times and places.

Certainty and Authority

The sentences that constitute my little linguistic experiment were invented, yet they might well have been collected from a place where two people were farming, laboring under a hot tropical sun. I suggest this because the difference between the context of farming and the context that is the location of the authority of a foundationist paper like my previous chapter matters. In part, the separation between them is remade in my foundationist argument. As well as separating words and materialities, my foundationist account also effects other divisions. It separates—as different categories—those that work with hoes and need to drink water from those who work with words and have no need of hoes or water, the author-in-the-text. While the former characters are burdened with responsibilities of many sorts, the latter characters are responsible only for reasoning in correct ways.

Foundationism seems to provide both a way for knowers to be certain and a way for them to be certain that they can be certain, the latter seeming to enable what Horton labeled as a commitment to "openness" over the first level of certainty. Certainty in knowing comes in learners, adopting, under instruction from nature or society, chains of reasoning through which they can know—logic. In foundationism, certainty about logic can originate only in recognized cognitive authorities who stand outside and above regular knowing with a privileged view of both structure and what is built with the structure—the foundation and its instantiations. We can think of certainty as flowing downward here in a system of legitimization. This downward flow of certainty, having cognitive authorities as the origins of certainty through their legitimization of cognizing structures, goes along with knowers as reasoning observers.

The version of foundationism I adopted insisted that many logics were possible as the result of different methods of symbolizing in representing, with certainty generated in the coherence of those representations. As I had it, this did not entail that there must be many worlds hermetically sealed off from one another by their different logics. According to the picture I presented, these logics constitute worlds of knowing that can be slipped into and out of by individual knowers; and exhaustive translations can be formulated and standardized. Explicitly, I insisted that users of one logic *can* explain it to another. Individual knowers can learn to work in multiple symbolic domains and formulate determinate methods of translation between the codes.

Nevertheless, these translations and accounts of logics can only be promulgated by experts: those with a privileged view of the general foundation categories created by past social practices, and of the particulars instantiated in differing cultural (symbolic) systems. Just as in universalist formulations, constituting certainty over logic is the preserve of cognitive authorities working with discursive strategies. My paper justifies and exemplifies the claims of its author to be such an expert. We can see how author(ity), located outside the domains both of the foundation and its representation, was constituted as embedded in the author-in-the-text, all invisibly accomplished in a literalization.

As such an author(ity), Quine opts for a grim despair in this, asserting that we have no alternative to banal translation. Imposition of our provincial linguistic habits onto the "other" is inevitable and justifiable, since "ours" (science's) is self-evidently the best we have. In contrast, in my relativist papers, I (author-in-text of chapters 3, 6, and 9) opt for a comical hope in locating categorization in shared yet disparate linguistic practices.[11] According to this hopeful formulation, "we" will always find ourselves already in the "other," as we find the "other" in us.

Quine and I prescribe alternative moral orders—worlds of despair or of hope—in dealing with difference. We prescribe answers to the question "How should we live?" These two limited and limiting responses make up the entire range of possible moral responses to the complex differences of our times and places from within a foundationist frame.[12] Universalists prescribe that "they" should all become like "us," with either a repellent triumphalism or an equally repellent despair: difference is abolished by declaration. Relativists insist on comical and elaborate harmonization by resolving difference in sameness—another way of abolishing difference by fiat. The only possible response of foundationism to difference is either pessimistic or optimistic denial. This limited response is precisely what I am now refusing.

Chapter Eleven

Embodied Certainty and
Predicating-Designating

*M*y overall project in this book is to reconfigure my relativist work on generalizing logics in ways that enable me to avoid the limitations of foundationist interpretations. My re-vision involves taking worlds as outcomes of collective going-on. I have already gone some way in figuring out quantifying generalizing within this new imaginary. My investigations of numbering and generalizing suggested that the objects they entail are generated in ritualized routines of performance in ordered/ordering microworlds. This implies that logics are embodied in collective going-on in specific times and places.

Reworking my past relativist studies, I developed the proposal that Yoruba life emerges through and with a logic of whole/part organization. In contrast, life in English language communities, and by extension science, emerges through and with a logic of one/many ordering. We need to take care in understanding this contrast; it is relational, and each style of reasoning necessarily implicates the other. One/many and whole/part orderings are versions of each other, particular ways of doing the relation unity/plurality.

Where does language fit into this? Language seems to hold a special place in questions of logic. The question "Is the structure of language caused by the structure of logic, or is it the other way around?" seems central. These contrasting causalities are captured as the two versions of foundationism. Universalists opt for the first configuration of cause and effect, relativists for the alternative. It is difficult to see language as anything other than representational, and in that sense explaining its role in logic is a major challenge for my imaginary of worlds as emergent. How can we explain language to fit with a notion of logic as carried in generative objects that are figured and in turn figure in the workings of ritualized routines in specific times and places? We need a new account of language for an imaginary of emergence to be credible as an interpretative frame. In particular, we need to reconsider the nature of predicating-designating; having it as a first step in a chain of logical representation is no longer credible. Developing that new account is the work of this chapter.

A clue to how to proceed here comes from my exploration of number names. In chapter 5, I suggested that in number names we see embedded an ancient ritual with fingers, toes, hands, and feet. An old routine has been shifted down, no longer embodied in performance with legs and arms, fingers and toes, but now as utterance. I argued that the ritual embedded in number names mimics a ritual embedded in doings with eyes and hands—quantifying generalizing. In chapter 8, I argued that quantifying generalizations, like volume, contain within them the figured/figuring materialities of quantifying microworlds. These proposals lead me to suggest that figured and figuring objects/subjects are emergent in predicating-designating. Talking reenacts a ritual that figures in particular ways.

Refusing Predicating-Designating as Growing from Removed, Judging Observation

One of the first problems to be overcome in developing a new account of language is how to understand predicating-designating in a way that does not have babies becoming judging observers, cognitively taking up a particular symbolic code of representation as they learn to talk. In foundationist understanding, predicating and designating is judgment. Language learners are necessarily removed observers, humans with powers of reasoning in minds located outside of, and making observations on, the material world. Universalist accounts have babies suddenly understanding the lineaments of material reality as they learn to talk. Relativists have them as suddenly understanding the postulates of language as they become predicators-designators. Foundationist formulations have a child learning to talk as struggling to become a willful knower. There, becoming a language user is the first step in becoming an individuated knowing human, capable of making further judgments about the world.

In my paper in chapter 9, I had children as judging observers cognizing the difference between "announcings of similarity" and "announcings of difference," and acting as reasoning humans in joining them together to predicate and designate. I had babies learning language as learning to classify through symbolic representation of action and object. I identified alternative possibilities for representing action/object with verbs/nouns. It is this incredible account of children's learning that I am specifically repudiating here.

Predicating-designating as judging in this sense has no place in the interpretive regime I am struggling for. Here we need to be able to understand predicating-designating and predicators-designators as specific outcomes, generated in the material histories of specific times and places. I am suggesting that the announcings that children produce as embodied acting are expressions of bodies in place. We could exaggerate and picture children's bodies as translating, as giving voice to place. A child's utterance grows from the grounded and robust certainty of her embodiment as participant. Here language use is embodied enacting, expression of and in turn expressing the material-symbolic orderings out of which it grows. Yet how can ordinary language users, speaking in grammatically correct ways, possibly be understood as *embodied* as predicators-designators? Bodies seem to be almost incidental to language use.

Outcomes of Predicating-Designating

I return to chapter 9 and my linguistic experiment. This time I tell things differently by embedding the sentences I contrived in settings. I put the words back into bodies, and times and places. Imagine a Yoruba academic visiting Australia. A colleague invites him to visit his home on the weekend. The Yoruba visitor arrives to find his host in his vegetable plot wielding an implement like a rake—a small blade on the end of a long thin pole. "What is this?" (Kí nìyìí?), he asks. "Ọkọ́ ni ó jẹ́" or "Ọkọ́ ni" ("It is a hoe" or "A hoe"), his host replies. Later, having introduced his guest to the family, he returns to his garden, and after some time the visitor appears with refreshments. "Kí nìyìí?" (What is this?), asks the gardener. "Omi ni ó jẹ́" or "Omi ni" ("It is water" or "Water"), his helpful guest replies. Similarly, we can imagine an anthropologist newly arrived in Nigeria, wanting to observe Yoruba farming practices. He accompanies a Yoruba farmer who has befriended him to his farm. The young anthropologist has never before seen the robust, short-handled, wide-bladed hoe of Yoruba farmers; he asks, "What is this?" (Kí nìyìí?) The farmer, tolerant and generous, answers, "Ọkọ́ ni ó jẹ́" or "Ọkọ́ ni" ("It is a hoe" or "A hoe"). The anthropologist makes his notes and later wanting to assist his sweating informant pours a cool drink from the vacuum flask he carries with him. The farmer straightens up and asks, "Kí nìyìí?" (What is this?). "Omi ni ó jẹ́" or "Omi ni" ("It is water" or "Water"), says the anthropologist, relieved that he remembers his Yoruba language classes.

Here we see that in sequences of sounds acts are presented, but it is not the act of the farmer/gardener wielding his hoe or the visitor pouring the cool water. They are specific acts of *being* by hoe and by water, and given what we already know of the workings of English and Yoruba numbers and generalizations, we might suspect that they specify the acts of being by hoe and water differently. I am suggesting that what emerges from "Omi ni ó jẹ́"/"It is water" and "Ọkọ́ ni ó jẹ́"/"It is a hoe" are water and hoes that are both at once multiple and singular, and the multiplicity is not only the contrast of long-handled, small-bladed implements with short-handled, wide-bladed ones. The sounds strung together thus embed rituals. Small rituals effect farmer/gardener and visitor as predicating-designating subjects, and hoe and water as objects, and we suspect that, just as subjects are multiple and singular, so too objects are multiple and singular. Those singularities/multiplicities are being managed by both men, and perhaps in different ways.

We might understand the Yoruba and English utterances as little rituals that interpellate the sounds, the water/hoe, and the men. This immediately raises two issues. A first question is "How did the men become embodied as predicators/designators?" It is routine acting for them as they respond to each other, but they were not born that way and perhaps for the first two years of their lives were not so embodied. How do children become embodied as interpellated by the ritual embedded in predicating-designating? I need to understand how predicating-designating is embodied acting for human bodies.

A second question lies in what might be taken as the flip side of that question. I need to understand how predicating-designating is embodied acting by material words. The puzzle here lies in identifying how the material, articulated sounds of the English and Yoruba utterances are meanings and meanings are sounds. How, for example, is an English-speaking child's "Puppy woof" already always predicating "woofing"—an action—of "a puppy"—always already a separate enduring entity? Equally, how are the sounds "Ajá tobi" already always "Dogmatter bigs"?

Formerly, I had predicating-designating as the product of minds working with the pure, given entities space, time, and matter. I had these resources unproblematically available to the mind of the developing cognizing child as available for categorizing manifestations of the material world to generate the categories spatiotemporal particulars and sortal particulars. In my re-vision, pure matter and empty space/time are outcomes, necessarily generated as objects in routines and repetitions, enacted within specific associations. Here I need to explain how subjects and objects are outcomes of predicating-designating, not the other way around.

I begin to deal with these puzzles by attending to acts, arguing that outcomes of acts are inherently relational. Next, picking up on recent work on theorizing the body by Brian O'Shaughnessy, I suggest that, through acting, children become embodied as certain of the relationalities generated in acts. I suggest that talk expresses this embodied certainty. However, expressing this certainty over relationalities in acts implies a figuring, necessarily a particular figuring. Referring to my conversations with Yoruba- and English-speaking children, I use some of their utterances again to explore the ways Yoruba speakers and English speakers figure the relationalities generated in acts.

At that point in the argument, I am part of the way in arguing for predicating-designating as figuring relationalities in acts, and explaining how Yoruba and English might differ in this. To complete my argument, I still need to show how bodily acts with arms and legs and utterances are one and the same. I do this by mobilizing an argument from the later Wittgenstein, where he has utterance as just a version of babies' prelinguistic responses to embodiment in a collective. Putting the pieces together, I show particular predicating-designating as specific figurings of relationalities generated in acts. It follows that certainty in this acting, like any other, originates in language users' embodied certainty of relationalities in acts. At the end we see that I have described a real difference between Yoruba and English in how the predicating-designating aspect of generalizing logic is done. We can recognize how it is possible to manage the multiple and singular hoes and water that the men manipulate with hand and arm movements and utterances, so that various connections and separations are contingent outcomes of their going on together.

Relationality in Acts

Consider the act of a toddler hauling himself up to a standing position. It is a slow and inexpertly achieved act; for these reasons, perhaps, it is more open to scrutiny. It seems like the individual act of an individual child. A closer examination shows it as collective. The table he hauls on and the floor he pushes against are also participants. Imagine him reaching up, grasping the edge of a small table or a low shelf, hauling with his arms, pushing with his feet and legs, looking for purchase with his toes. In the reaching of his arms, the straightening of his legs and back, we see extension. As he gradually assumes the perpendicular with respect to the floor, and perhaps feels himself tower over the table, the act plays out a relation.[1] I could say that his inexpert standing up is "a passage in space," but that would be too quick; we need to slow down to the pace of the baby. The extension (and I do not mean the muscular stretching of the baby's leg here) is the outcome of the interaction of embodied baby and embodied floor and table. The resistance of floor and table is as important as the accommodation of the arms and legs of the child in generating the extensions in and through which child's body, floor, and table are variably related across the act. Here/there is space, and it is generated as *relation* within the act itself. We could call it space, but we would be likely to forget its inherent relationality and begin to think of it as a given entity.

We could say also that the act is duration, a *relation* of a different sort, but again that might be too fast. The baby's standing has a before and an after. A then/now is, like here/there, an outcome of the collective act of standing. Both these relations are accomplished in the *relation* of accommodating legs and arms, and resisting floor and table through the act. In relation to the baby's bodily capacities, table and floor are generated as specific sorts of resistances. I want to point such extension, duration, and the specific resistances as *relations* that grow in acts. They are aspects of acts, aspects of each other. We can tell them, figure them, as separate and have them as outcomes of embodied collective acts. What I am doing here is imagining the setting of embodied baby with gradually strengthening arms and legs, floor, and table as a simple ordered/ordering microworld. Predicating-designating is a particular re-presentation of bodily acting, necessarily rending its complexity into complicated objects/subjects. The objects/subjects of sentences carry within them a particular figuring of acts.

Acts are both prefigured by, and promiscuously generative of, relations.

All the outcomes of acts are relational, and they are not produced as pure or as separate. It is not that (relational) extensions, durations, and resistances, just some of the outcomes of acts we might name, are generated as separated entities. All outcomes are profoundly implicated in all others, and they emerge as a whole.

I have been using a human, albeit a small one, as my focus because in this instance I am interested in how such embodied collective acts are implicated in language use. This should not lead us to equate acts with the conventional picture of acts as those of willful and reasoning humans. Acting is "going-on being," which rocks, trees, protozoa, hills, water, peanuts, policemen, parliaments, babies, floors, and tables all do alike in concert with myriad other actings, and mostly quite without human involvement. Human acts are just one source of orderings; acting on the part of nonhuman organisms and of the nonliving are equally sources of orderings in that they generate relational outcomes.

Embodiment as Certain of the Relationalities in Acts

I now extend the notion of acts as inherently generative of relations to argue that embodied certainty of the relationalities in acts is also an outcome of acting. I begin with a recent paper by Brian O'Shaughnessy, picking up on his concepts of "proprioception" and "long-term body image."[2] The new pictures O'Shaughnessy develops of how the matter of our bodies resides in modern space-time are quite a radical departure from accepted understandings. I excise his entities from his modern, causal interpretive frame, where separated, purely material things like bodies are identified as having particular inherent characteristics. Placing them within my interpretive frame, where human bodies as much as the subjectivities that are said to inhabit bodies are effects, I have "proprioception" and "long-term body image" as outcomes of collective acting. Here human bodies and their parts are not separable from what might be dubbed the "nonhuman-body bits of collective acting." I feel that this transplanting of O'Shaughnessy's concepts does them no violence but, on the contrary, makes his argument far more persuasive.

O'Shaughnessy develops a notion of body that transcends all the foundationist orthodoxies of bodies' insides and outsides. He begins by suggesting the possibility of a new sense, proprioception, which might be added to the famous five.[3] Conceding that the concept of a perceptual sense is vague and must be understood as radically dissimilar to other senses, like touch, in the end he effectively dismisses the notion of proprioception, and perhaps other sensitivities, as senses as they are conventionally understood in biology. The remarkable thing about this "sense" of proprioception is that nothing separates what we conventionally understand as the inside and the outside of the body. The body and nonbody fuse in action. O'Shaughnessy goes on to use this new sort of sense in developing his concept of "long-term body image." Here he begins by alerting us to the ambiguity in the general usage of the term "body image." He is very clear that he is avoiding the mentalistic overtones of its general usage; he effectively discredits sensation-representationalist theories. In consequence, his "long-term body image" has "a unique situation . . . and a very strange one at that: namely that the *revealed* (material object) constitutes the very system of ordering/individuation/differentiation of the *revealer* (bodily sensations). This property is a direct corollary of the radical immediacy of proprioceptive perception: the fact that proprioceptive sensations do not attentively mediate the perception of the object they help to make perceptible; for if they did, they would have an ordering system that was independent of their object."[4]

O'Shaughnessy's account brings with it a major problem. Having shown conclusively how the sensation-representationalist theory of proprioception is false, he is left with the pressing issue of where the extension (relational space) implicit in proprioception and eventually long-term body image comes from, since representing postural/bodily sensations is ruled out. This puzzle requires a drastic solution. His drastic solution is that the acting-body-in-place brings with it the content of extension. After considering my description of the baby and his microworld, we recognize that the relationality of extension will not be the only relation sedimented as knowledge in bodies, as outcome of acting. We can understand O'Shaughnessy's "long-term body image" as placed, bodily remembrance of performance of, among others, the relationalities of extension, duration, and variable resistances, all generated as relations within acts. Not only are extension, duration, and variable resistances all generated as relations within acts, but they are generated as bodies and with certainty. And once again, the relation and embodied certainty emerge as one.

Going on, O'Shaughnessy considers the origins of "long-term body image." Here the difference between my interpretive frame and his becomes evident. He isolates the memory, the image, in fingers themselves. For him, this embodied knowledge lies in the givenness of fingers, in the growingness of fingers, and in the incorporating possibilities of experience. I modify this understanding by putting the body back in place, taking body as emergent, not a given entity. As I see it, this is taking his notion of acting-body-in-place seriously. To locate body-in-place memory in fingers *as such* overlooks that what is important here, the suchness of fingers, lies in fingers in relation to the things they grip and squeeze. For example, while fingers-kneading-in-pastry-dough grow and sediment routines and repetitions in nerves, muscles, and bones, that is not the whole story. Fingers-kneading-in-pastry-dough take dough forms into forms of fingers. Unskilled kneading fingers become skilled kneading fingers. Equally, fingers are taken into the form of dough, and eventually pastry, which might be heavy or light. Fingers are the outcome of interactions with dough, and pencils, and so on, just as light or heavy pastry and neat or scrawled handwriting are outcomes of dough and pencils interacting with fingers. Bodies emerge as certain of the relationalities of their acting in place. Bodies are made as places (other versions or forms of), and places are made as (other versions or forms of) bodies.

Utterance as Outcome of Collective Acting

My contention is that this embodied certainty of relationality in acting that is an inevitable outcome of collective acting will, later in a child's life, take the form of language. However, I am only part of the way in making this argument. What I have yet to do is give an account of how acts are utterances and vice versa.

The assumption that language use originates in minds is widespread. I adopted this stance in my relativist paper in chapter 9. This sees the child as a detached observer of her world and suggests that a child uses labeling language when and only when she has no doubt that the present manifestation relates to past manifestations as continuity or difference. This implies that only when babies have satisfied themselves by further observations will they label manifestations. On this view, reasoning and doubt are present at the very beginning of our employment of language. Surely this is unwarranted aggrandizement, making something that is a matter of immediate expression of embodied being in a world of multiple embodied beings into something of reasoned intellect—the mind. It is a view of language that sees it as acting on the world from somewhere outside it, its categories derived from reasoned observation.

The account of language I am developing here differs in almost every way from this set of conventional understandings. In my new account, language use is part and parcel of embodied collective acting. As such, talking is another version of the certainty in relationalities generated in embodied acting. The objects and subjects of language are the outcome of predicating-designating, a particular figuring generated in what I have just told as an ordered/ordering microworld of acts. To complete my argument, I need to show how utterances, sequenced sounds, are an integral part of the fabric of embodied acting.

As I image things, sequences of sounds become meaningful sentences, working through and with other figurings circulating within a language community. This has babies learning to talk as first learning to imitate sound sequences, and then a gradual realization of predicating-designating as sound sequences constituted in and which constitute a figuring, as they get the pattern of the workings of sound sequences. Particular patterning of sounds is displayed and sometimes explicitly explained to children as they interact with older language users, figuring their emergent embodied certainty of the relationalities in acts.

The first element in this argument is to show how utterances, sequences of sounds, are part and parcel of embodied acting. In beginning, I piggyback on the later work of Wittgenstein. I opt in at the point where he argues that language use has its roots in babies' prelinguistic expressions of involvement with the material world.[5] Agreeing with Wittgenstein, I have babies' first words extending their practical involvement in and with the embodied world as they go on as embodied babies, but I have a problem with his exegesis in that he makes it seem that the baby is the only participant. As I see it, this negotiation is mutual accommodation and resistance on the part of the baby and the rest of the human and nonhuman participants in the baby's time and place.

Babies' first smiles at their parents and their first tracking behavior—following the mobile of red and yellow fluffy ducks with their eyes—are expressions of familiarity in embodiment. They are not matters of reasoning about the world. In the same way, babies' announcings or heralding utterances are expressions of familiarity of embodiment, the certainty of its relationalities. Just as grabbing a teething biscuit to bite on for relief is an expression, pulling at a sock and grunting and, months later, muttering "Osh osh!" ("Sock off!") while pulling at a sock are expressions of complex and mutual resistances and accommodations between baby embodiments and sock embodiments.

Wittgenstein presents us with an entirely different conception of language use, asserting its origins in certainty. He has the beginnings of language use as extension of physical actions stemming from familiarity with embodiment:[6] " 'Primitive' physical expression has no element of uncertainty, language use which grows out of physical expression begins in certainty."[7]

Suppose a child sitting in her high chair knocks over her cup of milk. Her prelinguistic response might be to pick up the cup and fling it to the floor—we would say she is expressing her anger—or perhaps it might be to point to the cup and cry out—expressing her distress. The child is in no doubt that the tipping of the cup and the mess of cold milk dripping off the table onto her legs are one and the same. She would not wait to observe what would happen in other cases; she just cries out and points, or reacts angrily, flinging the cup away, or perhaps if she is older shouts, announcing "Milk!" or "Spilt!" "There is a reaction which can be called 'reacting to the cause.'—We also speak of 'tracing' the cause; a simple case would be, say, following a string to see who is pulling it. If then I find him—how do I know that he, his pulling, is the cause of the string's moving? Do I establish this by a series of experiments?"[8]

The child acts immediately, flinging the cup, pointing, and crying; the person knitting immediately chases away the cat that has hold of the wool. There is no uncertainty, guessing, conjecturing, inferring, concluding. Emphasizing that this reacting is immediate highlights the aspect of acting implicit here. Words, articulated sounds, are an extension of, are no different from, these immediate reactions; words are an outcome of collective acting involving the milk and the cup as much as the child.

The flinging down of the cup is an immediate reaction, not touched by uncertainty. The "certainty" that Wittgenstein is talking of is certainty in embodied acting. A child's crying out with pain when injured is an occasion for her being taught to use the word "pain"; it does not presuppose that she has "the concept" of pain. Similarly, her first use of announcings like "Hot!" when she touches a flame, or "Puppy!" when a dog walks onto the scene, consists in coming to use them along with, or in place of, reactions like pulling her hand back or clinging to the nearest adult. It is not that announcings with words are somehow "mindful" or representational, while announcings with legs, torsos, and arms are bodily and quite separate. The point is that the acting is unified; wordy announcements are as much bodily as any other aspect of acting.

Predicating-Designating: A Ritual of Figuring Acts in Utterance

It is with some relief that I can now leave behind a preposterous understanding of babies learning to talk. No longer do I need to credit them with incredible powers of metaphysical awareness. No longer do I have them carrying out controlled experiments through inductive generalization about sameness and difference, generating objects and actions. Babies are left to go on being embodied acting babies, with the capacities of those acting bodies gradually increasing through that acting. This brings with it a new location for the certainty we associate with talking. Using language and being certain of it is no longer the preserve of a rational human mind. Talking is imbued with certainty because it grows as certain, just as walking is certain. We come to be certain we are doing it through being certain of the relationalities in it. Having embodied certainty of the relationalities in acts as an outcome of acting emerges as being certain we are walking here and now when we are walking. Uncertainty over acting would have bodies never moving purposively, including never uttering.

When a child has achieved in her utterances a soundscape with at least some variation, this soundscape becomes available for patterning or figuring. The little boy pushing with his legs against the floor and pulling with his arms on the table may utter a long grunt. The sound is the act. As he pulls at his sock, following his mother's example, an act is a sound sequence, "Osh osh!" gradually refined to "Sosh off!" This is recognizable to those around him (though not yet to him) as a predicating-designating utterance. An "off" is predicated of a "sosh." His accomplishment is celebrated by much repetition of his act by those around him. Over many repetitions, the ritualized acting will eventually wear the baby into doing verbs and nouns, doing predicating-designating.

In producing this utterance, there is no suggestion that the child has "read" the real structure of the world correctly, or suddenly become aware of the postulates of the language he is learning. He is going on as before, acting as embodied baby. His increasingly many multielement utterances emerge only gradually as figuring in the time and place of his being embodied baby.

A baby's doing predicating-designating clots as a routine. It is a ritual that, emerging in her embodied acting, in time comes to interpellate producer of the sounds as figurer of relations in acts. Predicating-designating is figuring relationalities generated in acts, learned as ritualized acting.

We can understand predicating-designating as ritualized performance, carried out within utterances, of the collective act of figuring relationalities in acts. It interpellates language users as figuring relations in acts. Grammar is a way of doing some of the relationalities in acts.

Differences in Figuring Relationalities Generated in Acts

This brings me to differences in predicating-designating among those babies who grow in Yoruba-speaking communities and those nurtured in English-speaking communities. In chapter 6, I described how I had spent many hours listening to English-speaking and Yoruba-speaking children talk about a little performance I made with cups, glasses, water, and peanuts. The ready answers the children gave to my questions showed that in both language communities stories about how we figure relationalities in acts are readily told and apparently circulate freely. Recruiting water in glasses, peanuts in bowls, and tins as my coactors, I acted out little performances and asked children to comment on that (collective) act. Almost all of them readily did so. This capacity to articulate the figuring in acts is not esoteric and abstract knowledge. These girls and boys were certain of their story. I understand their certainty as arising in their embodied certainty of the relationalities in acts.

Here I translate again my previous translation of the quotations I gave of children's statements, using the terms I have developed to describe outcomes of acts: relational extensions, durations, and resistances. The children I spoke to (chapter 6) told me how to "see" the water and the peanuts correctly so as to make a generalization. In chapter 8, I retold what they said, to have them telling the acts of being of water and peanuts in terms of stuff enclosed and its boundaries. I said English-speaking children said that you need to watch the enclosed stuff, whereas Yoruba-speaking children told me you had to watch how the stuff was enclosed.

Speaking of how to figure the acts of the water, peanuts, and so on, keeping tabs on the *stuff* implies that what I should foreground is the resistances that water and peanuts put up against the accommodations of my glasses/cups and fingers/hand. It is important, said the English-speaking children, to watch out for the relation of resistance/accommodation in the act: the push-pull of water in a glass against the push-pull of the walls of the glass and of my fingers and hands grasping and manipulating the glass. In figuring the relationalities in acts, an English-speaking child has acts as involving relational resistances—what we ordinarily call matter. Necessarily backgrounded are the relationalities duration and extension. What emerges as objects in English language predicating-designating are specific durations-extensions—what we might call "space-time bits." The specific durations-extensions are the being of matter. That being of matter is the subject in English sentences. (It is difficult to say this without sounding quite daffy.)

Yoruba-speaking children gave me quite a different story about figuring acts. They said that you must watch out for the movements, the extensions—the here/theres in the acts themselves, the hereness and the thereness, and importantly, the sequencing of the there/then and the here/now, the relative enduring. They told me that I needed to watch how the peanuts were being divided between here and there, what was being done then, in relation to what is being done now. To paraphrase: these helpful Yoruba children said that the most important thing about acts, the being in the act, is extension and duration as relations in the act. The *relation* that Yoruba speakers like 'Dupe have as the subjects of sentences is resistances—what we might call "stuff of this sort or that." Because figuring acts in Yoruba predicating-designating has as significant the doing of here/now–then/there, resistances emerge as what the here/now–then/there is done in and as.

We are back where we started in the first section of chapter 9. Yet not quite; the long diversion has freed us from the modern myth that space, time, and matter are given and must be represented. We now understand them as just a particular way of telling outcomes of acting. We are now free to use them without the elaborate charade of foundation and instantiation. The words "matter," "space," and "time" are *not* representing now, they are recognizable as expressions of being human. They are another form of the relationalities that are the outcomes of acts—figured material objects generated in the ritual of predicating-designating.

The idea that words are expressions of acts involved in being human is of course far from a novel idea. It is perhaps how most people feel about the languages they speak. It is also an assumption of many academic disciplines—those that generally call themselves the humanities, and differentiate themselves from the sciences. Science, claiming a different sort of certainty, one justified from the outside, so to speak, insists on language as representation and finds itself compromised by the literalizations it depends on and, in consequence, boxed-in in its efforts to understand itself in new ways. Perhaps it is time for science to leave behind its hall of mirrors; surely it no longer has need of it.

Predicating-designating has reemerged in a different location from that I elaborated in my relativist paper, and so has the difference between Yoruba and English. Predicating-designating are doings within the body of collective acting, just another transition, another form of transactional generalizing. The objects and subjects emergent in predicating-designating have the same sorts of features as those emergent in other forms of generalizing. It is another way of managing complexity by coming up with entities that carry their relationalities within. The subjects and predicates of our sentences are entities that participate in collective going-on through mobilizing the associations implicit in those relations.

So where is the difference between English and Yoruba predicating and designating? Asking this is tied up with questions that ask about creolization and pidgins. Being banal, everyday affairs, these rituals of predicating-designating can be influenced and changed; they can hybridize. New rituals that mesh more or less, in this way or that, with old ones can emerge and do the job well enough, in speaking this language or that. New banal predicating-designating rituals can emerge in acting together so as to go on in some fashion. In the process of creolization, new grammars might be generated. Thus in one sense, the difference in predicating-designating lies in the pidgin in which much of my everyday life in Nigeria was lived. This too is not a new idea.

One/Many and Whole/Part

Yoruba predicating-designating sets relationalities in acts in a way that it seems to have acts as done by a whole—a sort of matter—which appears as a part here and now. English figures relationalities in acts another way. It seems to take acts as starting with a part—a space-time bit—that is constitutive of a whole—a sort of matter. I am suggesting that the figuring in English and Yoruba predicating-designating meshes with figurings I have already identified in the ways generalizing and numbering are done. Yet we need to take care here.

We have seen one rather similar type of figuring in contemporary Yoruba life and another as pervasive in the life of English language communities and, by extension, science. Respectively, a whole/part figuring and a one/many figuring. In doing predicating-designating, doing quantifying generalizing, and doing numbering in quite distinct types of little rituals, the whole/part and the one/many figurings echo and allude to each other. Yet it is important to recognize that, while they might link up in loose ways, none of the three forms of ritualizing microworlds stands in a determining position with respect to the others.

In emergent contemporary worlds, figurings circulate and sometimes link up and reinforce each other. Particular relationalities are encountered over and again. Just as figurings can link up, they can also interfere and interrupt. Sometimes these interruptions cause disconcertment. The figurings implicit in Yoruba language talk interrupt the figurings implicit in doing generalizing and numbering in scientific ways, and what emerges sometimes is a disconcerting hybrid ritual.

In understanding the working of the contrast whole/part and one/many, perhaps the first point that needs to be made is that the contrast whole/part–one/many is *relational*. Reasoning by doing either necessarily involves proceeding to the other. Reasoning is the cycling of each into the other. Each element of the contrast implies the other. In pointing to difference, I am pointing to a contrast in ways of beginning reasoning in Yoruba language life and English language life, and, by extension, science. In projects involving both—like my students' science lessons—the difference is likely to emerge as the issue of how we begin our dealings, and how we might go on so to encourage the emergence of new versions of the objects we are dealing with, in the form of making new rituals in our microworlds.

The whole/part–one/many contrast is also present in foundationist reasoning. There the first is called deduction, which is contrasted to the latter, induction. In foundationism, the cycling of whole/part–one/many, deduction-induction, is seen as constituting a necessary ascent in abstraction. The cycling enables an ascent to certainty through increasing abstraction.

In emergent worlds constituted by the odd new sorts of objects I have been showing, we cannot see whole/part as deduction and one/many as induction, since we have abandoned the separated domains of the general and the particular. Now we have only one emergent complexity, contingently managed. The manys and ones, and the wholes and their parts are linked in messy and contingent ways—all categories are outcomes. What is a one and what is a many, what is a whole and what is a part, is an outcome of the doing, a linking of this object with its internal relationalities, to that object, in this way or that.

In my new stories of doing numbers, generalizing, and talking in English and Yoruba, I have struggled to stay true to the disconcertments I experienced in Yoruba classrooms many years ago. Having reached this point, I ask myself whether I would do things differently in laboratory classes teaching mathematics and science to teachers, were I to find myself back in Ile-Ife in Nigeria. Accepting that the resourcing of Nigeria's primary school classrooms may now be even more meager, would I encourage a new cohort of students to do things in ways that differ from what I had my students doing in the mid-1980s, just before I left Nigeria? Probably I would not. My years among my Yoruba friends changed me, and I mean changed in a bodily sense. I learned to trust my students' classes and to trust them as teachers, and their pupils as learners. Encouraging my students to do science and mathematics lessons in practical ways, bringing to the fore the actual doing of the little rituals of quantifying with hands, eyes, water, string, and rulers as well as with utterances turned out to be a useful and generative way to deal with the tensions between English and Yoruba generalizing logics of numbering.

Letting these little rituals happen as they would, which is what I eventually learned to do, is to trust in an embodied certainty of the time and place of the lessons themselves. I learned to trust teachers, and to trust myself to know when lessons were successful and when they were not. This trust derived from criteria found in the doing of those lessons, not from criteria found in the pages of curriculum documents, or in philosophy books on the nature of numbers. Learning when to tell stories about numbers and, more importantly, what to tell when, and how to link them with other stories suitable for other occasions, I learned to trust in an embodied certainty of time and place. That is not only trusting people as subjects figured in particular ways, often ways that are unfamiliar. It is also learning to trust objects figured in unfamiliar ways.

Postcolonial Sciences?

A familiar object "generalizing logic" has been reborn in this book. It has emerged as both multiple and singular. There is a strong possibility that it will never emerge from this book into further projects. It might, however, be picked up by, say, a group of environmental scientists and social scientists attempting to grapple with the fraught issues of how to maintain biological diversity and develop more productive Yoruba farming practices.[9]

In such a project as that, this strange, newly reborn object could make a difference. We can imagine this strange multiple object as participating in an endeavor to develop some responses in Nigeria and elsewhere, to the apparent clash between the need to prevent "biodiversity loss" and the need to contrive ways in which farm plots might produce more food more efficiently.

One effect of having this new object participate in such a project would be to drastically multiply the forms or versions of biodiversity. As we have seen, this does not mean sacrificing its singularity, but it does mean being canny in managing the singularity/multiplicity. We would be able to recognize the many ways that biodiversity can be done both by Yoruba farmers and by scientists. Both Yoruba farmers and environmental scientists work ritualizing ordered/ordering microworlds to do biodiversity. One of the differences is that the former can be expected to work their farms (or what it is that is a whole—which may well not be a farm[10]) through the whole/part relation, while the latter work their field sites as a one/many.

If we respect the differences and contrive links between the objects, new routines might be invented in environmental science that effect a quite new version of the object "biodiversity." This biodiversity could be more intimately linked to place, perhaps. That is what beginning with a whole/part reasoning could contribute here, and the issue of place is a significant problem for the scientific versions of biodiversity that we currently have. Linked with various old Yoruba farming versions of "diversity," and with the already existing multiple scientific versions of "diversity," hybrid forms—the working of old microworlds in new ways and inventing new sorts of microworlds—could emerge. In such a project, a newly reborn, more complicated object "biodiversity" would be radically strengthened by becoming more robust and plastic, its multiplicity enhanced in useful ways. I suspect that some Yoruba environmental scientists might already be working in this way. Perhaps my stories might help them name what they are doing in new and useful ways.

The at once multiple and singular object generalizing logics, born in this book, might take on life in ways of doing biodiversity and telling about that doing. If that happened, we would see scientists doing ontological politics, doing a politics of difference as they manage difference in generalizing logics by contingently making connections and separations.

Implicit in the stories I have told as the third chapter in each part of my book has been the view that science should give up its old and tired commitment to representationism or foundationism. Actually, I have a sense that already that change is well under way in many parts of science, and we might hope that philosophers and popular science writers (many of whom are also scientists) will soon catch up. I am suggesting that science that gives a chance for doing a politics of difference mobilizes generalizing logic as transitional, as what transactions might be accomplished in.

Having logics as transitional/transactional requires a minimalist set of assumptions about what "we" the collective share. What is implicit in this is, first, a notion of worlds as generated in acting in collectives. Second, it assumes that all collectives have an embodied certainty of the relationalities in acting.

In this science, we might invent new ordered/ordering microworlds, or new rituals and routines to do in our old ones. Here science will tell itself as managing complexity through inventing new sorts of ritualizing microworlds. Telling such stories would be telling of the rituals and the coparticipants, human and nonhuman, living and nonliving, in micro-worlds, as reliable ways of managing complexity. In the past, these sorts of stories have largely been confined to telling of the emergence of sub-jects—and have often been labeled anthropology. I am extending that sort of storytelling to accounts of the emergence and lives of objects. In doing this, I recognize that one thing that those of us who make these studies need to learn to do better is to tell our studies in ways that have the objects emerging as recognizable and usable, without giving rise to the impression that the worlds we are dealing with are somehow not real. The threats to so many of our times and places are so pressing, we simply cannot afford the luxury of so-called science wars.

Notes ✧

Chapter One

1. African Primary Science Program Materials, Education Development Centre, Newton, MA, 1970.

2. Teaching was not my first choice of a career. Despite a Ph.D. and prestigious connections in metabolic biochemistry (in the shape of Sir Hans Krebs as an examiner of my thesis, and a paper in the *Biochemical Journal*), by my late twenties I found myself, with two babies, unemployable as a research biochemist. Revising career plans, I completed a diploma in primary education and took up teaching kindergarten children, eventually making a career as a teacher educator in science. It was in this capacity that I took up my lecturing position at Obafemi Awolowo University in Nigeria in 1979. That position turned out to enable yet another career change into science studies in 1987, after my return to Australia.

The time and place out of which this book grew, was a moment of exhilarating hope in Nigeria's history. My Nigerian colleagues expressed a confidence that, through dedicated work in Nigeria's robust and lively universities, new futures could be invented. The moment had been ushered in partly with the establishment, after many years of military rule, of a civilian political regime in 1979 as I arrived in Ile-Ife with my family. It also grew from a determination, on the part of many ordinary Nigerians, for reconciliation in the aftermath of the horror of what many know as the Biafran war. Yet it was a moment that was lost in the continuation of a political culture that thrived, and perhaps still thrives, in generating and regenerating a limiting and brutalizing conformity.

See Wole Soyinka, *The Open Sore of a Continent: A Personal Narrative of the Nigerian Crisis* (Oxford: Oxford University Press, 1996); and Michael Watts, "Nature as Artifice and Artifact," in *Remaking Reality: Nature at the Millennium* (London: Routledge, 1997), 243–68.

3. Mr. Ojo and the other teachers I tell of are fictional characters. I have abided by the convention of changing details in telling of them and their pupils. In some cases, the episodes I tell of are composites of several incidents; in others, they allude to a specific occasion.

4. Michael Taussig, *Mimesis and Alterity: A Particular History of the Senses* (New

York: Routledge, 1993), 225. The gasp or the laughter at such moments is a familiar refrain in postcolonial cultural studies. Michael Taussig assembles several instances and quotes Benjamin: "The perception of similarity is in every case bound to an instantaneous flash. It slips past, can possibly be regained, but really cannot be held fast" (Walter Benjamin, "Doctrine of the Similar," *New German Critique* 17 [spring 1979]: 66).

5. Taussig, *Mimesis and Alterity*, 224–25. Similarly, the *molas* displayed in "The Art of Being Kuna," an exhibition at UCLA's Fowler Museum of Cultural History, 1997–98, are breath-catching in their hilarious subversion of American breakfast-cereal logos.

6. Taussig, *Mimesis and Alterity*, 129.

7. Bruno Latour, *We Have Never Been Modern*, trans. Catherine Porter (Cambridge: Harvard University Press, 1993), 91–92: "I proposed anthropology as a model for describing our world [but] I quickly recognised that this model was not readily useable, since it did not apply to science and technology. . . . If anthropology is to become comparative . . . it must be made symmetrical. To this end, it must become capable of confronting . . . the true knowledge to which we adhere totally. It must therefore be made capable of studying the sciences by surpassing the limits of the sociology of knowledge and, above all, of epistemology."

8. H. J. Reed and Jean Lave, "Arithmetic as a Tool for Investigating Relations between Culture and Cognition," *American Ethnologist* 6 (1979): 568–82.

9. Jean Lave, "The Values of Quantification," in *Power, Action, and Belief: A New Sociology of Knowledge*, ed. John Law, Sociological Review Monograph 32 (London: Routledge and Kegan Paul, 1986), 88–111.

10. T. N. Carraher, D. W. Carraher, and A. D. Schliemann, "Mathematics in the Streets and in the Schools," *British Journal of Developmental Psychology* 3 (1985): 21–29.

11. Lave, "Values of Quantification," 88.

12. The quote comes from Barbara Lloyd, "Cognitive Development, Education, and Social Mobility," in *Universals of Human Thought: Some African Evidence*, ed. Barbara Lloyd and John Gay (Cambridge: Cambridge University Press, 1981), 186, but she had shown "the delay" in Yoruba children in 1971 in "Studies in Conservation with Yoruba Children of Differing Ages and Experience," *Child Development* 42 (1971): 415–28.

13. Lloyd, "Cognitive Development, Education, and Social Mobility," 184.

14. "The greatest gap in the Yoruba's knowledge and control of natural phenomena is the lack of any technique for discriminating between the essential and unessential qualities of things; between what constitutes a thing as such and what are merely accidental properties" (N. A. Fadipe, *The Sociology of the Yoruba* [Ibadan: University of Ibadan Press, 1970], 293). Fadipe is mobilizing an old European "mechanical philosophy" as expounded, for example, by Descartes. This explained all physical effects as resulting from the size, shape, and motion of insensible bits of matter. These were to be distinguished from secondary effects such as color and texture, which are imposed by the mind of the beholder.

15. For example, see Rudolf Carnap, *Philosophical Foundations of Physics: An In-*

troduction to the Philosophy of Science, ed. Martin Gardner (Chicago: University of Chicago Press, 1966).

16. C. R. Hallpike, *The Foundations of Primitive Thought* (Oxford: Clarendon Press, 1979), 252.

17. Lloyd, "Cognitive Development, Education, and Social Mobility," 182. The work with Tiv children was reported by Rhys Price-Williams, "A Study concerning Concepts of Conservation of Quantities among Primitive Children," *Acta Psychologica* 18 (1961): 297.

18. Hallpike, *Foundations of Primitive Thought,* 252.

19. Kathryn Pyne Addelson, "Knowers/Doers and Their Moral Problems," in *Feminist Epistemologies,* ed. Linda Alcoff and Elizabeth Potter (New York: Routledge, 1993), 265–94. Pyne Addelson uses the notion of "public problem" as a unit in her feminist epistemology. For her, it is "the process of making moral problems public [and] one of the central processes through which dominance is maintained—and changed." She explicitly differentiates her notion of "public problem" from what she calls the common sense or folk understanding, which she believes is shared by many philosophers. She rejects a notion of public problem "which has something like the following format: A social condition objectively exists (scientists confirm it) that is problematic because it has morally relevant consequences. The problematic situation comes about through people doing various things for various reasons or causes. The solution lies in understanding and changing the conditions (either people's behaviour or the environment) so that the future will be different (i.e., the social problem will be solved or resolved)" (276–77).

20. Following the trail of texts, we can identify 1967 as a high point in that old debate. In the developing philosophical debate on "African thought," Robin Horton's 1967 "African Traditional Religion and Western Science" (*Africa* 37:50–71, 155–87) was widely recognized as pivotal. Horton took a Popperian line on the analysis of scientific rationality and extended it to "traditional African religion," purporting to show the "openness" of science as against the "closedness" of "traditional African religion." "Open" and "closed" are terms Popper uses in elaborating his particular brand of positivistic scientific rationality (*Conjectures and Refutations: The Growth of Scientific Knowledge* [New York: Basic Books, 1962]. Rationality had become the hot topic in philosophy of science following the publication of Kuhn's *Structure of Scientific Revolutions* (Chicago: University of Chicago Press, 1962). Perhaps we can credit Winch with the introduction of Africans into this debate with his "Understanding a Primitive Society" (*American Philosophical Quarterly* 1 [1964]: 307–24). In 1970, the collection *Rationality* (ed. Bryan Wilson [Oxford: Basil Blackwell, 1970]) appeared as a contribution to debate in philosophy of science. Horton continued his approach in his paper in the collection he edited with Ruth Finnegan, *Modes of Thought* (Faber: London, 1973), and philosophers at the University of Ife continued to critique Horton's analyses (Barry Hallen, "Robin Horton on Critical Philosophy and Traditional Religion," *Second Order* 6, no. 1 [1977]: 81–92; Barry Hallen and J. O. Sodipo, *Knowledge, Belief, and Witchcraft* [London: Ethnographica, 1986]). Anthropologists Geertz and Gellner had their say (Clifford Geertz, *The Interpretation of Cultures* [New York: Basic Books, 1973]; Ernest Gellner, *Legitimation of Belief* [Cambridge: Cambridge University Press, 1974]). Kwasi Wiredu

was the first African voice to be clearly and widely heard (*Philosophy and an African Culture* [Cambridge: Cambridge University Press, 1980]). By 1982, when *Rationality and Relativism* (ed. M. Hollis and S. Lukes [Oxford: Basil Blackwell, 1982]), which includes a rethink from Robin Horton, was published, the debate had largely puffed itself out.

On the cross-cultural psychology side, Gay and Cole published their *New Mathematics and an Old Culture* (New York: Holt, Rinehart and Winston, 1967), which reported their work with Kpelle people of Liberia in West Africa, following hot on the heels of the 1966 publication of *Studies in Cognitive Growth,* edited by the well-known psychologist Jerome Bruner among others, in which the content was strongly cross-cultural (J. S. Bruner, R. R. Oliver, and P. M. Greenfield, eds., *Studies in Cognitive Growth* [New York: Wiley, 1966]). We can trace the continuation of this work in M. Cole and S. Scribner, *Culture and Thought: A Psychological Introduction* (New York: Wiley, 1974) and the collection edited by J. W. Berry and P. R. Dasen, *Culture and Cognition: Readings in Cross-Cultural Psychology* (London: Methuen, 1974), and recognize Hallpike's *Foundations of Primitive Thought* (1979) as both a definitive work and in some senses a last gasp in this phase of the debate. I have lumped apples and oranges together here. These are rather differently located traditions of psychology. However, they are all alike in juxtaposing "tradition" and "the modern" in telling modern stories of modernity.

21. There is currently a reappraisal of logical positivism (or as some now term it logical empiricism) in philosophy of science. While many now see this project, which flourished in Europe, the United States, and Australasia between the 1930s and the 1960s, as fatally flawed, it continues to exert considerable influence in philosophy of science, if only in the sense that many philosophers are still struggling to free themselves of its epistemological tentacles. See, for example, Ronald Giere and Alan Richardson, eds., *Minnesota Studies in the Philosophy of Science,* vol. 16, *Origins of Logical Empiricism* (Minneapolis: University of Minnesota Press, 1996).

22. Carnap, *Philosophical Foundations of Physics,* 52.

23. Soyinka, *Open Sore of a Continent,* 75. As it happened, my head of department, apparently acting in an excess of enthusiasm for the War against Indiscipline, for a time refused to sign my contract renewal, citing the fact that my students taught science and mathematics in Yoruba as proof of my incompetence. It seemed for some time that our family might need to leave Nigeria with great haste. He weakened only after considerable pressure from more powerful faculty.

Chapter Two

1. See chapter 3, "Numerals in Language Use."

2. With this characterization of my study, I do not mean to imply that these papers actually had an impact, that they changed things for any African children or their teachers, or even had any impact among psychologists, who for the most part are now quite uninterested in cross-cultural issues. One significant exception is Michael Cole; see his *Cultural Psychology: A Once and Future Discipline* (Cambridge: Harvard University Press, Belknap Press, 1996).

By the time the papers were published, the realist-relativist debate that characterized philosophy of science and various social sciences in the 1960s and 1970s had largely lost interest in "Africa" and its "thought." Development discourse remained in its comfortable universalism. The papers created some gratifying excitement and interest among constructivist theorists of knowledge, mostly for the promise they seemed to hold of establishing the credentials of relativism. As I show in this chapter, this promise cannot be fulfilled.

3. I am indebted here to Bruno Latour, "A Few Steps toward an Anthropology of the Iconoclastic Gesture," *Science in Context* 10 (1997): 63–83. Latour's paper was reworked to become chapter 9, "The Slight Surprise of Action Facts, Fetishes, and Factishes," in *Pandora's Hope: Essays on the Reality of Science Studies* (Cambridge: Harvard University Press, 1999).

We might locate both my old studies, and this new treatment of them, in a cluster of studies that claim to be something more than ethnomathematics, where there is "the presumption that the concept of numbers and the procedures of arithmetic are unaffected by cultural differences" (Gary Urton, *The Social Life of Numbers* [Austin: University of Texas Press, 1997], 7). In this group of studies that go beyond ethnomathematics, I would include Urton's elegant study, along with the equally remarkable *Intimations of Infinity* by Jadran Mimica and *Signifying Nothing: The Semiotics of Zero* by Brian Rotman. These studies "place everything on the table—numbers, arithmetic, philosophy, and mathematics—not privileging any particular form of human knowledge and practice as being above either language or culture" (Urton, *Social Life of Numbers*, 8).

My old studies of Yoruba number share an iconoclastic intent with those of Urton, Rotman, and Mimica. Putting this in a polite formulation, Urton suggests that such studies "hold out to mathematics the offer of expanding and clarifying its philosophical grounding, thereby enriching both mathematics and anthropology" (19). The flaws in my old study that I identify and elaborate in this chapter are also relevant to the studies of Urton, Rotman and Mimica. My contention in this chapter is that the framing our studies share prevents us from holding anything out to mathematics, or any other group of scholars.

4. David Bloor, *Knowledge and Social Imagery,* 2nd edition (Chicago: University of Chicago Press, 1991).

5. Lorraine Code has considered one aspect of this at length in *Epistemic Responsibility* (Hanover, NH: University Press of New England, 1987) through critiquing the limitedness of the notion of propositional knowledge. She takes up the issue afresh in *Rhetorical Spaces: Essays on Gendered Locations* (New York: Routledge, 1995, 22), emphasizing that the questions about epistemic responsibility are multiple and "not uniform in type or in provenance. Some of them focus on the construction of knowledge, others on its deployment and dissemination, some more on process, others more on product. Some bear on the credibility of enquirers, their interests in the enquiry, what they stand to lose or gain in power or prestige. Others bear upon the willingness of such enquirers to subject their most cherished conclusions and commitments to critical scrutiny: upon what they are prepared to let go or reexamine in the interests of truth and justice. Some epistemic responsibility questions are about how knowledge is put to use and its social-institutional effects. . . . It generates a complex system of questions that are mutually catalytic in

keeping lines of critique open, and issues of accountability always, potentially, on the agenda."

Kathryn Pyne Addelson (*Moral Passages: Towards a Collectivist Moral Theory* [New York: Routledge, 1994]) is more explicit, pointing to the "double participation" of researchers, a notion she picked up from Prudence Rains (*Becoming an Unwed Mother: A Sociological Account* [Chicago: Aldine Atherton, 1971]). She sees responsibility as lying in the participations of researchers as much outside the academy as inside. "Today, at the end of the twentieth century, what is the place of academic professionals? What ought it to be? What is scholarly freedom today and what collective possibilities are in the offing and for whom? . . . What is it to do morally and intellectually responsible work? . . . There are no answers to these questions if we stick to definitions of responsibilities under the old folk concept of the academic professions. . . . The question of responsibility must be answered by asking, What is our freedom? What might our collective possibilities become? What moral passage might we embark upon? These are open questions that have to be answered in practice, as we professional knowledge makers do our work reflexively so that we may in the end do it responsibly. But the process cannot begin without an intimate knowledge of who the 'we's' ought to be, and who is to make the collective possibilities and for whom. . . . That of course is the question of how we should live" (182).

6. Pyne Addelson, *Moral Passages,* 182.

7. See A. O. Lovejoy, *The Great Chain of Being* (Cambridge: Harvard University Press, 1936); A. O. Lovejoy, *Essays in the History of Ideas* (Baltimore: Johns Hopkins Press, 1948).

8. In a recent publication, I explore origins of these assumptions (Helen Verran, "Logics and Mathematics: Challenges Arising in Working across Cultures," in *Mathematics across Cultures: The History of Non-Western Mathematics,* ed. Helaine Selin and Ubiritan D'Ambrosio [Dordrecht: Kluwer Academic, 2000], 55–78).

9. In his famous paper ("African Traditional Thought and Western Science," *Africa* 37 [1967]: 50–71, 155–87), Robin Horton ascribes a form of fetishizing to "traditional African thought." He has the "systems of thought" as "closed," implying that "words (concepts) are stuck fast to things" so that the system of thought cannot progress. See chapter 10 for some discussion.

10. I am using the Latourian phrase here, *We Have Never Been Modern,* 92.

11. Latour, "Slight Surprise of Action."

12. Taussig, *Mimesis and Alterity,* 225.

13. Michael Carter, "Notes on Imagination, Fantasy, and the Imaginary," in *Imaginary Materials: A Seminar with Michael Carter,* ed. John Macarthur (Brisbane: Institute of Modern Art, 2000), 67. I am grateful to Helen Smith for drawing my attention to Carter's work.

14. Cornelius Castoriadis, *The Imaginary Institution of Society* (Cambridge: MIT Press, 1987), 26.

15. See, for example, B. Barnes, *Interests and the Growth of Knowledge* (London: Routledge and Kegan Paul, 1977).

16. This way of understanding the term "postcolonial" links up with Stuart Hall's observation that the term points to "a notion of a shift or a transition conceptualised as a reconfiguration of a field, rather than as a movement of a linear transcendence between two mutually exclusive states. Such transformations are not only not completed but they not be best captured within a paradigm which assumes that all major historical shifts are driven by a necessitarian logic towards a teleological end" ("When Was the 'Post-Colonial'? Thinking at the Limit," in *The Post-Colonial Question: Common Skies, Divided Horizons,* ed. Iain Chambers and Lidia Curti [London: Routledge, 1996], 254).

17. Marilyn Strathern, *The Gender of the Gift: Problems with Women and Problems with Society in Melanesia* (Berkeley: University of California Press, 1988); Marilyn Strathern, *After Nature: English Kinship in the Late Twentieth Century* (Cambridge: Cambridge University Press, 1992); Marilyn Strathern, *Reproducing the Future: Anthropology, Kinship, and the New Reproductive Technologies* (New York: Routledge, 1992).

18. Strathern, *Reproducing the Future.*

19. Marilyn Strathern, "Environments Within: An Ethnographic Commentary on Scale," Linacre lectures, 1996–97, Linacre College, Oxford, 1.

20. Marilyn Strathern, "The Decomposition of an Event," *Cultural Anthropology* 7, no. 7 (1992): 244–54. I am grateful to Katayoun Sadghi Rad Hassall for drawing my attention to the significance of this article.

21. Strathern, "Decomposition of an Event," 245, 249.

22. Karin Barber, *I Could Speak until Tomorrow: Oriki, Women, and the Past in a Yoruba Town* (Edinburgh: Edinburgh University Press, 1991).

23. Toyin Falola, "In Praise of *Oriki,*" *Journal of African History* 33 (1992): 331–32.

24. "*Oriki* are a master discourse. They are composed for innumerable subjects of all types, human, animal, and spiritual; and they are performed in numerous modes or genres. They are compact and evocative, enigmatic and arresting formulations, utterances which are believed to capture the essential qualities of their subjects, and by being uttered, to evoke them. They establish unique identities and at the same time make relationships between beings. They are a central component of almost every significant ceremonial in the life of the compound and town; and are also constantly in the air as greetings, congratulations and jokes. They are deeply cherished by their owners" (Barber, *I Could Speak until Tomorrow,* 1).

25. Barber, *I Could Speak until Tomorrow,* 26.

26. "The exegesis of *oriki* reaches its fullest and most institutionalized form in the genre called *itan. Oriki* and *itan* are separate but symbiotic traditions. . . . While women are the principal carriers of *oriki,* it is usually old men who tell *itan.* And while *oriki* are continually in the air . . . *itan* are rarely told except when occasioned by a family dispute, a chieftaincy contest or a direct request for information from a son in the compound (or, of course, a researcher)" (Barber, *I Could Speak until Tomorrow,* 28).

27. Barber, *I Could Speak until Tomorrow,* 30.

28. Barber, *I Could Speak until Tomorrow*, 31.

29. Strathern, "Decomposition of an Event," 245.

30. Barber, *I Could Speak until Tomorrow*, 290.

31. Michel Callon, John Law, and Arie Rip, eds., *Mapping the Dynamics of Science and Technology: Sociology of Science in the Real World* (London: Macmillan, 1986); Michel Callon, "Some Elements of a Sociology of Translation: Domestication of the Scallops and the Fishermen of St Brieuc Bay," in *Power, Action, and Belief: A New Sociology for Knowledge?* ed. John Law, Sociological Review Monograph 32 (London: Routledge and Kegan Paul, 1986); Michel Callon, "Techno-economic Networks and Irreversibility," in *A Sociology of Monsters: Essays on Power, Technology, and Domination,* ed. John Law, Sociological Review Monograph 38 (London: Routledge, 1991); Michel Callon, "Actor-Network Theory: The Market Test," in *Actor Network Theory and After,* ed. John Law and John Hassard (Oxford: Blackwell, 1999).

32. Bruno Latour and Steve Woolgar, *Laboratory Life: The Construction of Scientific Facts* (Beverly Hills: Sage Publications, 1979); Bruno Latour, *Science in Action: How to Follow Scientists and Engineers through Society* (Milton Keynes, United Kingdom: Open University Press, 1987); Bruno Latour, *The Pasteurization of France,* part 1 (Cambridge: Harvard University Press, 1988); Bruno Latour, "The 'Pédofil' of Boa Vista," *Common Knowledge* 4 (1995): 144–87; Bruno Latour, "From the World of Science to the World of Research," *Science* 280 (April 1998): 207–8; Bruno Latour, "On Recalling ANT," in *Actor Network Theory and After,* ed. John Law and John Hassard (Oxford: Blackwell, 1999), 15–25.

33. Callon, Law, and Rip, *Mapping the Dynamics of Science and Technology;* John Law, "On the Methods of Long Distance Control: Vessels, Navigation, and the Portuguese Route to India," in *Power, Action, and Belief: A New Sociology for Knowledge?* ed. John Law, Sociological Review Monograph 32 (London: Routledge and Kegan Paul, 1986), 234–63; John Law, introduction to *A Sociology of Monsters: Essays on Power, Technology, and Domination,* ed. John Law, Sociological Review Monograph 38 (London: Routledge, 1991); John law and Michel Callon, "The Life and Death of an Aircraft: A Network Analysis of Technical Change," in *Shaping Technology—Building Society: Studies in Sociotechnical Change,* ed. Weibe Bijker and John Law (Cambridge: MIT Press, 1992), 21–52; John Law, *Organising Modernity* (Oxford: Blackwell, 1994); John Law, "After ANT: Complexity, Naming, and Topology," in *Actor Network Theory and After,* ed. John Law and John Hassard (Oxford: Blackwell, 1999), 1–14.

34. Michel Serres, *Angels: A Modern Myth,* ed. Philippa Hurd (Paris: Flammarion, 1995); Michel Serres, *Detachment,* trans. Genevieve James and Raymond Federman (Athens: Ohio University Press, 1989); Michel Serres, *Genesis,* trans. Genevieve James and James Nielson (Ann Arbor: University of Michigan Press, ca. 1995); Michel Serres, *Hermes: Literature, Science, Philosophy,* ed. Josue V. Harari and David F. Bell (Baltimore: Johns Hopkins University Press, 1982).

35. Annemarie Mol and Marc Berg, "Principles and Practice of Medicine: The Co-existence of Various Anaemias," *Culture, Medicine, and Psychiatry* 18 (1994): 247–65; Annemarie Mol, "Missing Links, Making Links: The Performance of Some

Atheroscleroses," in *Differences in Medicine: Unravelling Practices, Techniques, and Bodies,* ed. Marc Berg and Annemarie Mol (Durham: Duke University Press, 1998), 144–65; Annemarie Mol and John Law, "Situated Bodies and Distributed Selves: On Doing Hypoglycaemia," paper presented to Netherlands Graduate School of Science, Technology, and Modern Culture/Centre de Sociologie de l'Innovation Conference on the Body, Paris, September 1998; Annemarie Mol, "Ontological Politics: A Word and Some Questions," in *Actor Network Theory and After,* ed. John Law and John Hassard (Oxford: Blackwell, 1999), 74–89.

36. I read the table that begins on page 219 as doing this, for example (Donna Haraway, *Modest_Witness@Second_Millennium.FemaleMan©_Meets_OncoMouse™* (New York: Routledge, 1997).

37. Pyne Addelson, "Knowers/Doers and Their Moral Problems"; Pyne Addelson, *Moral Passages.* Symbolic interactionism, a form of qualitative sociology, might be understood as the empirical arm of American pragmatism. Here, too, worlds emerge all of a piece, yet the framing retains a foundationism by having researchers as participant observers on the one hand and theorists on the other. Pyne Addelson significantly alters the symbolic-interactionist frame by abandoning this duality.

38. "Being intellectually responsible, requires, for professionals like myself, devising theories and practices that can make it explicit what the collective activity is and what some important outcomes of the activity might be" (Pyne Addelson, *Moral Passages,* 18).

39. Latour, *We Have Never Been Modern,* 103.

40. I first heard Zoe Sofoulis put the slogan this way at a workshop held in the Department of History and Philosophy of Science at University of Melbourne, June 1996.

41. Latour, *We Have Never Been Modern,* 11.

42. Latour, *We Have Never Been Modern,* 134.

43. Latour, *We Have Never Been Modern.*

44. Marilyn Strathern, "Afterword," in *Shifting Contexts: Transformations in Anthropological Knowledge,* ed. Marilyn Strathern (New York: Routledge, 1995), 178.

Chapter Three

1. Samuel Adjai Crowther (later Anglican bishop), *Vocabulary of the Yoruba Language* (London: Church Missionary Society, 1843), and the American missionary T. J. Bowen, *Grammar and Dictionary of the Yoruba Language* (Washington, DC: Smithsonian Institution, 1858).

2. Adolphus Mann, "Notes on the Numeral System of the Yoruba Nation," *Journal of the Royal Anthropological Institute* 16 (1887): 60, 62.

3. Samuel Johnson, *The History of the Yorubas from the Earliest Times to the Beginning of the British Protectorate* (Lagos: CMS [Nigeria] Bookshops, 1921); R. G. Armstrong, *Yoruba Numerals* (Ibadan: Oxford University Press, 1962); R. C. Abraham, *Dictionary of Modern Yoruba* (London: Hodder and Stoughton, 1962).

4. An early version formed part of the paper published as Helen Watson, "Applying Numbers to Nature: A Comparative View in English and Yoruba," *Jour-*

nal of Cultures and Ideas 2–3 (1986): 1–19. This fuller version was never published; it formed part of a manuscript "Numbers and Things," completed in 1989, which purported to be a relativist account of logic in quantifying. I did not try to publish the manuscript, such was my ambivalence about it. The way I had erased from the text the difference that I sought to grasp and myself as a bumbling participant in contemporary Yoruba life, making myself a disembodied authority, had produced a text that I felt contributed to a continuing colonizing agenda. I needed quite a different way to understand numbers and things. Struggling to work that out, I then realized (again and again) that I was unable to write that way of understanding numbers and things. I needed far more than words. This way of proceeding—presenting my old texts and going on to dissolve the boundaries they depend on—is a compromise that I feel accomplishes some of what needs to be accomplished, and perhaps manages to resist at least to some degree the conflation of authoring and authorizing.

5. This account of English language numeration has been compiled from Tobias Dantzig, *Number: The Language of Science* (New York: Free Press, 1954); C. L. Barber, *The Story of English* (London: Pan Books, 1972); Karl Menninger, *Number Words and Number Symbols: A Cultural History of Numbers* (Cambridge: MIT Press, 1969); Georges Ifrah, *From One to Zero: A Universal History of Numbers* (New York: Viking, 1985).

There are remnants of many recursive name sequences in English that use different bases: bases twelve and twenty, echoing the troy system of measuring weight, were embedded in English and Australian currency until thirty years ago. We still see the duodecimal connection when we buy eggs in dozens; base twenty (score) was also common until relatively recently.

6. Barber, *Story of English*, 80.

7. Dantzig, *Number*, 12.

8. Dantzig (*Number*, 32) places this move in Europe at the end of the thirteenth century.

9. My main sources here are Johnson, *History of the Yorubas;* Armstrong, *Yoruba Numerals;* and most importantly Abraham, *Dictionary of Modern Yoruba.*

10. There are four sets of numerals in Yoruba usage. The list of number names in table 3.1 and the derivations shown in table 3.2 use the group of numerals known conventionally as the cardinal set. These differ from the primary set I have already presented, by incorporating the verb *mún* (to collect). In developing these tables, I might also have used the form of the numerals that incorporates the term for cowrie *(owó)*. For two reasons, I am of the opinion now that these tables should be constructed with that latter set of numerals. First, the set incorporating the term *owó* set has become the norm for children learning the number names, although the connection with cowries seems to have been more or less forgotten. Second, strictly speaking, the set of names I have adopted here should not be used without naming what it is that has been collected together. When I wrote this chapter some ten years ago, I assumed that this was the set I should work with because, as I understood things then, this was the cardinal set. I now feel that this set should be termed the multiplicity set, and consider that the notion of cardinality has no relevance to Yoruba numbers.

11. S. A. Ekundayo, "Vigesimal Numeral Derivational Morphology: Yoruba Grammatical Competence Epitomized," paper presented in the Linguistics Department, University of Ife, 1975.

12. Fadipe, *Sociology of the Yoruba.*

13. Ayo Bamgbose, private communication; Ayo Bamgbose, *A Short Yoruba Grammar,* 2nd edition (Ibadan: Heinemann, 1974); P. Oladele Awobuluyi, *Essentials of Yoruba Grammar* (Ibadan: Oxford University Press, 1978).

14. Abraham, *Dictionary of Modern Yoruba.*

15. Abraham, *Dictionary of Modern Yoruba.*

16. Bamgbose, *Short Yoruba Grammar;* Awobuluyi, *Essentials of Yoruba Grammar.*

17. In commenting on this section in my original manuscript "Numbers and Things," Karin Barber remarked that "[t]he numbers associated with counting currency seem actually to be the *first* ones that many Yoruba children learn. I suspect that the connection with cowries has more or less been forgotten, and that *oókan, eéjì, ẹẹta* . . . is now interchangeable with *ení, èjì, ẹ̀ta. . . .* What I heard them chanting all day long in Òkukù was this: "ONE-*ookan!* TWO-*eéjì!* THREE-*ẹẹta!* They were learning English language, English counting, and Yoruba counting all at once."

18. Ifrah, *From One to Zero,* 433.

19. Dantzig, *Number,* 12.

20. It seems to me that attempts to identify the referents of number words are as misguided as attempts to identify the referent of a part of a ruler. What is important about natural number names is that they form a sequence. Within any numbering system, it is the place marked and how this is achieved that matters.

The contrast between von Neumann's and Zermelo's numbers that I make here derives from the formalist project to elaborate the foundations of mathematics. The formulations are ways of exemplifying the different algebras of von Neumann and Zermelo.

Through the nineteenth century and into the twentieth, conceptions about the objects and aims of mathematics were changing. The axiomatic method of constructing mathematics on set theoretic foundations, where every mathematical theory is the study of some algebraic system, was gradually taking shape. This project holds that mathematical theory is the study of a set with distinguished relations, in particular algebraic operations, satisfying some predetermined conditions or axioms. From this point of view, all number systems (natural number, real number, complex number, and so on) are algebraic systems. In elaborating this for the system of natural number, one encyclopedia of mathematics (M. Hazewinkel, ed., *Encyclopaedia of Mathematics* [Dordrecht: Kluwer Academic, 1993], 497) gives an account that can be understood as an expression of the algebras of J. von Neumann, introduced into mathematics in the 1930s. This is the conventional way of expressing this move: "by the system of natural numbers one usually understands the algebraic system $N \langle N, +, \cdot, 1 \rangle$ with two algebraic operations: addition ($+$) and multiplication ($\cdot$), and a distinguished element (1) (unity), satisfying the following [8] axioms."

Nevertheless, using the Zermelo theorem, a different account of natural number is possible. Elaborated some thirty years earlier, Zermelo's theorem says that every set can be well-ordered such that every normal subset of S has a first element (one preceding all others of the subset). Zermelo proved that every set can be well-ordered if it is assumed that one element of any subset can be chosen (or designated) as a "special" element. This assumption is called the axiom of choice, or the Zermelo axiom.

21. Commenting on this in my original manuscript in 1986, Karin Barber noted, "Zero—the word used for this in schools (e.g., for 'nought out of ten') is òdo, and there is a slang word for dunce, olódo. I guess this word (i.e., òdo) was made up to fill the gap (literally), but have no idea of its etymology. Àìwà also exists (I think more commonly) as àìsí (because the negative of ó wà is kò sí), e.g., in oríkì I have heard 'Àìsí Àjàyí nílé, ìlù tóró' (Ajayi's not being at home, the town was quiet)." Robert Armstrong used odo as his added zero.

22. Armstrong, *Yoruba Numerals*, 21.

23. Ayo Bamgbose, personal communication; Bamgbose, *Short Yoruba Grammar*; Awobuluyi, *Essentials of Yoruba Grammar*.

24. Abraham, *Dictionary of Modern Yoruba*.

25. Hallpike, *Foundations of Primitive Thought*.

26. For example, see Carnap, *Philosophical Foundations of Physics*.

27. Bloor, *Knowledge and Social Imagery*.

Chapter Four

1. Literalizing features prominently in the discussion of chapters 4, 7, and 11. I find a recent consideration of literalism by Vincent Crapanzano (*Serving the Word: Literalism in America from the Pulpit to the Bench* [New York: New Press, 2000]) useful in defining how I use the term. Crapanzano locates literalism in a position midway between essentialism and iconic representation. For Crapanzano, essentialism is arguable and has analytic depth; he finds it difficult to imagine thinking that is fully free from essentialism. Iconic representation, on the other hand, has immediacy and translatability, both absent in essentialism. In iconic representation, one image can quickly substitute for another, whereas essentialism suggests fixity.

Crapanzano is deeply interested in literalism, and he is troubled by confusion between iconic representation, literalism, and essentialism. As he sees it, the "focus in literalism is on the relationship between word and meaning rather than any essential feature of the meaning carried by the word. The relationship is simply asserted: it is neither analytically derived nor based iconically on similarity. . . . [Literalism] resists the easy translation characteristic of iconic representation . . . [but] it is in figurative language that danger lurks for the literalist. . . . he or she is destined to uphold a moral, spiritual, legal (or other) world order by preserving and serving the literal word" (353–54). That description of literalism seems to fit precisely my relativist studies. I found myself upholding a particular world order by preserving and serving some words. Chapters 4, 7, and 11 tell of how I came to recognize that.

2. E. W. Said, *Orientalism* (New York: Vintage Books, 1978), 72, quoted in Arjun Appadurai, "Number in the Colonial Imagination," in *Orientalism and the Postcolonial Predicament: Perspectives on South Asia*, ed. Carol Breckenridge and Peter van der Veer (Philadelphia: University of Pennsylvania Press, 1993), 314–40.

3. This form of "accounting" is just as important in the accumulating that has been recognized as a democratic safeguard against elitism. Theodore Porter comments on this duality in numbers as technology in considering political philosophy's assessment of numbering and quantifying: some have numbers as the instrument of domination, others equally passionately hold quantifying as a tool of freedom (*Trust in Numbers: The Pursuit of Objectivity in Science and Public Life* [Princeton: Princeton University Press, 1995], 73).

4. Appadurai, "Number in the Colonial Imagination," 317.

5. See David Turnbull, *Maps Are Territories: Science Is an Atlas: A Portfolio of Exhibits* (Chicago: University of Chicago Press, 1993).

6. Appadurai, "Number in the Colonial Imagination," 316–17.

7. See Nicholas Thomas, *Entangled Objects: Exchange, Material Culture, and Colonialism in the Pacific* (Cambridge: Harvard University Press, 1991), 134.

8. "Even in the case of the Aboriginal Tasmanians, where written and visual representations of people bestialize or convey ugliness in an offensive and profoundly racist manner, artifacts are depicted in the standard way" (Thomas, *Entangled Objects,* 133).

9. We need to think of fingers and hands in place. Fingers and hands emerging from the serge jacket of the British Museum attendant differ from those grasping the fan during a ceremony in a Maori community.

10. J. D. Y. Peel, "The Cultural Work of Yoruba Ethnogenesis," in *History and Ethnicity,* ed. Elizabeth Tonkin, Maryon McDonald, and Malcolm Chapman (London: Routledge, 1989), 198–215.

11. Ian Hacking, "Language, Truth, and Reason," in *Rationality and Relativism,* ed. Martin Hollis and Steven Lukes (Cambridge: MIT Press, 1982).

12. Nancy Stepan, *The Idea of Race in Science: Great Britain, 1800–1960* (London: Macmillan, 1982).

13. Mann, "Notes on the Numeral System of the Yoruba Nation," 60.

14. C. H. Toy, "The Yoruban Language," *Transactions of the American Philological Association* 9 (1878): 19.

15. Toy, "Yoruban Language," 19.

16. Linguists were not the only participants in rendering languages as standardized objects. For example, J. D. Y. Peel says of Yoruba, "If one wants a precise time and place for the inauguration of the modern Yoruba language, it would be in 1844, when Samuel Ajayi Crowther opened the first Yoruba service in Freetown. . . . the language itself, whose orthography took another thirty years to settle, was thoroughly hybrid: its morpho/syntax predominantly Oyo/Ibadan, its phonemes markedly of Abeokuta, its lexicon enriched by coinings and the speech of Lagos and Yoruba disapora" ("Cultural Work of Yoruba Ethnogenesis," 202).

17. Menninger, *Number Words and Number Symbols;* Ifrah, *From One to Zero;*

T. Crump, *The Anthropology of Numbers* (Cambridge: Cambridge University Press, 1990); C. Zavlavsky, *Africa Counts: Number and Pattern in African Culture* (Boston: Prindle, Weber, and Schmidt, 1973).

18. See P. J. Davis and R. Hersh, *The Mathematical Experience* (London: Pelican Books, 1983); and Crump, *Anthropology of Numbers.*

19. Levi Leonard Conant, *The Number Concept: Its Origins and Development* (New York: Macmillan and Co., 1896), 33: "Our perfect system of numeration enables us to express without difficulty any desired number."

20. Conant, *Number Concept,* 4.

21. "The variety in practical methods of numeration observed among savage races, and among civilized peoples as well is so great that any detailed account of them is impossible. [A vast range of notches, knots, splints, etc.] have been and still are to be found in the daily habit of great numbers of Indian, negro, Mongolian and Malay tribes; while to pass at a single step to the other extremity of intellectual development, the German student keeps his beer score by chalk marks on the table or wall. . . . [But] the German student scores his reckoning with chalk marks because he might otherwise forget; while the Adaman Islander counts on his fingers because he has no other method of counting,—or, in other words, of grasping the idea of number" (Conant, *Number Concept,* 8).

22. Conant, *Number Concept,* 94. From his description, the Japanese numerals seem just as opaque as any of the other systems he presents, so the basis of the approval is not clear at all.

23. Using sanitized language of "a typology of number uses" and an "ecology of number" in his recently published *Anthropology of Numbers* (1990), Crump, for example, still finds "the Far East is par excellence the home of numerate cultures" (ix), whereas Africans are presented as being more interested in affairs of the spirit. His enterprise is similar to Conant's. He is looking at "the ecology" of the way the universal domain of numbering enters as particular instances in various societies. He is trying to catalogue the different sorts of entangling, you might say. But he finds it difficult to escape the hierarchies of those whose work he uses. There are better and worse ways to represent the domain, and there is inherent worth in representing, and hence in having good representations of, the domain.

24. This is an incorrect translation of Yoruba to English.

25. Conant, *Number Concept,* 70–71; the quote is from R. F. Burton, "Notes on Certain Matters Connected with the Dahoman," *Memoirs Read before the Anthropological Society of London* 1 (1865): 314, but Conant notes Burton made a miscalculation and had the Yoruba or the Dahomans—it's not clear which—losing 6,820 men.

26. Conant *(Number Concept)* cites Burton, "Notes on Certain Matters," 314.

27. Conant, *Number Concept,* 32–33.

28. Conant, *Number Concept,* 85, 48.

29. Thomas, *Entangled Objects,* 127.

30. Thomas, *Entangled Objects,* 128.

31. J. W. Burrow, *The Crisis of Reason: European Thought, 1848–1914* (New Haven: Yale University Press, 2000).

32. This brief (!) intellectual history of the twentieth century follows Latour, *We Have Never Been Modern*.

33. For example, Johnson participated in the negotiations through which a settlement in "the Yoruba wars" was achieved (Peel, "Cultural Work of Yoruba Ethnogenesis").

34. For example, in talking of the Ijeshas, John Peel writes, "Though the Yoruba clearly fall well within the Yoruba ethnographic category and share the legends of Ife dynastic origins, they only came to consider themselves 'Yoruba' in the course of the twentieth century. Taking on a name once borne by one of their historic enemies. . . . In fact the Ijesha adoption of the Yoruba identity, though apparently 'traditional,' has proceeded parallel to the adoption of other, obviously 'modern,' identities as Nigerians, Christians or Muslims" (*Ijeshas and Nigerians: The Incorporation of a Yoruba Kingdom, 1890s–1970s* [Cambridge: Cambridge University Press, 1983], 15).

35. Johnson, *History of the Yorubas*, xix.

36. Johnson, *History of the Yorubas*, xxxiii.

37. Johnson, *History of the Yorubas*, xxxvii.

38. Johnson, *History of the Yorubas*, 1.

39. Ekundayo, "Vigesimal Numeral Derivational Morphology." I have been unable to find a published version of this paper; I have a mimeographed version.

40. S. A. Ekundayo, "Aspects of Underlying Representations in the Yoruba Noun Phrase," Ph.D. thesis, University of Edinburgh, 1972.

41. Ekundayo, "Vigesimal Numeral Derivational Morphology," 2.

42. Appadurai, "Number in the Colonial Imagination."

43. Douglas Hofstadter, *Gödel, Escher, Bach: An Eternal Golden Braid* (Harmondsworth: Penguin Books, 1980), 142.

44. I am grateful to Cass Wigley for helping me recognize that this social movement around recursion had been influential in my study.

45. Sherry Turkle, *The Second Self: Computers and the Human Spirit* (New York: Simon and Schuster, 1984), 317.

46. Turkle, *Second Self*, 304.

47. R. Audi, ed., *The Cambridge Dictionary of Philosophy* (Cambridge: Cambridge University Press, 1995); T. Honderich, ed., *Oxford Companion to Philosophy* (Oxford: Oxford University Press, 1995).

48. Eric S. Raymond, *The New Hacker's Dictionary*, foreword and cartoons by Guy L. Steele, Jr. (Cambridge: MIT Press, 1993).

49. Kurt Gödel, *On Formally Undecidable Propositions* (New York: Basic Books, 1962). This is an English translation of Gödel's 1931 paper.

50. Burrow, *Crisis of Reason*.

51. Turkle, *Second Self*, 219.

Chapter Five

1. A. Akinyele to district officer in Ibadan, 18 August 1921, Bishop A. B. Akinyele Papers, Maps and Manuscripts Collection, Kenneth Dike Memorial Library, University of Ibadan, Box 10, Duplicates of Letters, Letter Book April 1921–January 1922. I am indebted to Ruth Watson for passing this letter on to me.

2. In 1921, Ibadan's population was estimated at 238,075, compared to Lagos's at 99,690.

3. Ruth Watson, *"Civil Disorder Is the Disease of Ibadan": Chieftaincy and Civic Culture in a Colonial City* (London: James Currey, 2001), chapter 5.

4. Watson, *"Civil Disorder Is the Disease of Ibadan,"* chapter 5.

5. In a letter to Grace Grier, District Officer M. S. Grier writes, "[Ibadan] is a difficult place for there is no hereditary chief as in Oyo where they believe in the divine right of Kings and the result [in Ibadan] is continual intrigue and want of proper authority; consequently chaos" (Manuscripts [Africa] Collection, 1379 Grier 2/8f.8, Kenneth Dike Memorial Library, University of Ibadan). This quotation was kindly supplied by Ruth Watson.

6. Watson, *"Civil Disorder Is the Disease of Ibadan,"* chapter 5.

7. Bolanle Awe, "The Ajele System: A Study of Ibadan Imperialism in the Nineteenth Century," *Journal of the Historical Society of Nigeria* 3 (1964): 47–71.

8. Watson, *"Civil Disorder Is the Disease of Ibadan,"* chapter 5.

9. Watson, *"Civil Disorder Is the Disease of Ibadan,"* chapter 5.

10. Watson, *"Civil Disorder Is the Disease of Ibadan,"* chapter 5.

11. Strathern says of the relation, "It is *holographi* in the sense of being an example of the field it occupies, every part containing information about the whole and information about the whole being enfolded in each part" (*Relation Issues in Complexity and Scale,* 17). She credits Bohm and Wagner, *Symbols That Stand for Themselves* (Chicago: Chicago University Press, 1986).

12. Strathern, *Relation Issues in Complexity and Scale,* 18–20.

13. Louis Althusser, *Lenin and Philosophy and Other Essays,* trans. Ben Brewster (New York: Monthly Review Press, 1971), 170–83. This notion is currently enjoying a renaissance. On some occasions, Donna Haraway uses the term in ways that are very close to Althusser's use (*Modest_Witness@Second_Millennium.Female Man©_Meets_OncoMouse™,* for example, pages 50 and 58), and on others it takes on the wider and fuller allusions with which I have taken up. Judith Butler also makes use of the term in "Conscience Doth Make Subjects of Us All," in *The Psychic Life of Power: Theories in Subjection* (Stanford: Stanford University Press, 1997), and *Excitable Speech: A Politics of the Performative* (New York: Routledge, 1997).

14. In this sense, we can understand that Callon's intéressement and recruitment (see Callon, "Some Elements of a Sociology of Translation") are an example of interpellation.

15. We might understand collective acting as constituted by myriad ideologies. A major distinction between ideology and collective acting is the understanding that ideology has concrete forms that are realized in ideological state apparatuses (Althusser, *Lenin and Philosophy and Other Essays,* 181). Some of the ideologies that

constitute collective acting are certainly so enshrined—that is Donna Haraway's point throughout *Modest_Witness@Second_Millennium.FemaleMan©_Meets_Onco Mouse™*.

16. Althusser, *Lenin and Philosophy and Other Essays,* 172.

17. In addressing his reader directly, Althusser says, "I wish to point out that you and I are *always already* subjects, and as such constantly practice the rituals of ideological recognition, which guarantee for us that we are indeed concrete, individual, distinguishable, and (naturally) irreplaceable subjects. The writing I am currently executing and the reading you are currently performing are also in this respect rituals of ideological recognition" (*Lenin and Philosophy and Other Essays,* 172). And this of course applies equally well to me and my readers and the collective acting we are participants in (subject to and subjects of). In a footnote to the above passage, Althusser notes that the double "currently" in the passage illustrates the way ideology is eternal. Since I can appropriate his passage written on 6 April 1969 and make it serve for me by rewriting it on 7 October 1997 and my readers can read it on any subsequent day, his point is doubly made.

18. Mol and Berg, "Principles and Practice of Medicine"; Mol, "Ontological Politics"; Mol, "Missing Links, Making Links"; see also Mol and Law, "Situated Bodies and Distributed Selves."

19. Mol, "Ontological Politics," 79.

20. Mol, "Missing Links, Making Links," 161–62.

21. Latour's "Do Scientific Objects Have a History? Pasteur and Whitehead in a Bath of Lactic Acid" (*Common Knowledge* 5 [1996]: 85) charts the "transition, rarely studied, between two completely different ontological states."

22. We could imagine such a history of the enumerated entity as analogous to the history of lactic yeast, which Latour tells in "Do Scientific Objects Have a History?" 70–89.

23. G. J. Afolabi Ojo, *Yoruba Culture: A Geographical Analysis* (London: University of London Press, 1966), 131.

24. Afolabi Ojo cites a colonial official document here: M. Perham, *Native Administration in Nigeria* (London, 1937), 162.

25. Afolabi Ojo, *Yoruba Culture,* 119–20.

26. Mol, "Ontological Politics," 74.

Chapter Six

1. Gay and Cole, *New Mathematics and an Old Culture,* 1.

2. Cole and Scribner, *Culture and Thought,* 172.

3. Lloyd, "Studies in Conservation with Yoruba Children," 415–28; Barbara Lloyd and Brian Easton, "The Intellectual Development of Yoruba Children," *Journal of Cross Cultural Psychology* 8 (1977): 3–15; Lloyd, "Cognitive Development, Education, and Social Mobility," 176–96; P. M. Greenfield, "On Culture and Conservation," in *Studies in Cognitive Growth,* ed. J. S. Bruner, R. R. Oliver, and P. M. Greenfield (New York: Wiley, 1966), 225–56.

4. Lloyd, "Cognitive Development, Education, and Social Mobility," 186.

5. Hallpike, *Foundations of Primitive Thought.*

6. Lloyd, "Studies in Conservation with Yoruba Children," 427.

7. Frank Musgrove, *Education and Anthropology: Other Cultures and the Teacher* (Chichester: John Wiley and Sons, 1982).

8. Helen Watson, "Learning to Apply Numbers to Nature: A Comparison of English Speaking and Yoruba Speaking Children Learning to Quantify," *Educational Studies in Mathematics* 18 (1987): 339–57. A grant from the Research Grants Committee of the University of Ife made this investigation possible.

9. The format of the puzzles was the common Piagetian "conservation" tasks, but the theoretical framework of the study is not Piagetian. I do not share the view of universalist number that underlies Piaget's work. I take a Wittgensteinian approach to number and assert that it is not necessary to postulate numbers as extralinguistic objects. Numbers are not entities beyond what is written down or said. Numbers provide a linguistic machinery for making detailed and precise observations and descriptions. They arise in symbolic transformations.

10. An extended family might consist of three generations living in the same compound. Typically, these might be a man and his wives and children, together with various members of his parents' generation. Children of a compound tend to constitute a society that in many ways keeps its affairs separate from adult society. The older children fulfill the roles of caregivers and teachers for the younger ones.

11. I am grateful for the cooperation and interest of the children and staff of St. Patricks School, Toro, Oyo State, Nigeria; the Obafemi Awolowo University Staff School, Oyo State, Nigeria; and Flemington Primary School, Victoria, Australia.

12. 'Funmi Oyekanmi and 'Nike Akinola interviewed the Yoruba-speaking children in my presence, and 'Lola Durojaiye transcribed and translated the Yoruba language interviews. Their skilled help is gratefully acknowledged.

13. "Thingness" is a neologism; the orthodox way of naming this quality is "numerosity." This quality is the basis of counting. To begin counting, we have to make a unit—a thing. In counting, we report the extent of the quality of numerosity or thingness (how many things) in this collection. To allocate the quality of thingness or numerosity is to qualify, just like using an adjective before a noun is to qualify.

14. In a relativist account of quantifying, however, qualities are understood as categories generated in the history of quantifying by the antecedents of English-speaking communities.

15. Mr. Laosebikan, lecturer in mathematics education at the Institute of Education, Obafemi Awolowo University, explained the two sorts of units in Yoruba measurement as ẹyọ (ìdì), a bundled up or contrived unit like a basket of kola nuts or a hand span that cap makers often use in measuring their pieces of cloth, and ẹyọ (ohun), a unit found as such, like a single kola nut or a cap.

16. To grasp the distinction between *mode* and *quality*, it can help to remember English grammar lessons: adjectives qualify, adverbs modify.

17. Gottlob Frege, *The Foundations of Arithmetic*, trans. J. L. Austin (Oxford: Basil Blackwell, 1974), 61e.

18. W. V. O. Quine, *Word and Object* (Cambridge: MIT Press, 1960).

19. See Helen Watson, "Investigating the Social Foundations of Mathematics: Natural Number in Culturally Diverse Forms of Life," *Social Studies of Science* 20 (1990): 283–312; also chapter 9.

20. Hallpike, *Foundations of Primitive Thought,* is a relatively recent exponent of this view, but it is a position that informed much of psychological investigations of African children. See chapter 1, note 20.

21. There are more sophisticated versions of this than the explication by Carnap, Quine's account, for example, which I deal with in chapter 10.

Chapter Seven

1. Latour, *Science in Action,* 227.

2. I am taking up the notion of microworld from Joseph Rouse (*Knowledge and Power: Towards a Political Philosophy of Science* (Ithaca, NY: Cornell University Press, 1987). Rouse adopted the term from Marvin Minsky and Seymour Papert (MIT Artificial Intelligence Laboratory Memo, no. 2999, September 1973), considerably extending it from their notion of microworlds as "isolated fictional domains whose objects and possible events can be represented in a computer program" (Rouse, 105) to link it up with Foucault's sites of interrogation, normalization, and tracking. Rouse confines his use of microworlds to laboratories; I am extending it to include not only the laboratory-like field sites of my experiment, but also the most banal and ordinary sites of getting on in collective life.

3. Ian Hacking, *Representing and Intervening: Introductory Topics in the Philosophy of Science* (Cambridge: Cambridge University Press, 1983), 220. Hacking and Rouse share the view that laboratories (microworlds) are constitutive of phenomena generated by ordering complexity in routine practices.

4. Rouse, *Knowledge and Power,* 101.

5. Whitehead (*Process and Reality: An Essay in Cosmology* [New York: Free Press, 1957], 96) is keen to rebut the "baseless metaphysical doctrine of 'undifferentiated endurance,'" which has its origin in a mistake about objects like stones for example. He argues that "for Descartes and for commonsense . . . the immediate percept . . . the quiet undifferentiated endurance of the material stone . . . *is* the stone. [It] is a basic notion which dominates language, and haunts both science and philosophy." To go beyond an everyday use of this notion is an error for the scientist or philosopher. "The simple notion of an enduring substance sustaining persistent qualities, either essentially or accidentally, expresses a useful abstract for many purposes of life. But whenever we try to use it as a fundamental statement of the nature of things, it proves itself mistaken. It arose from a mistake and has never succeeded in any of its applications. But it has had one success: it has entrenched itself in language, in Aristotelian logic, and in metaphysics. For its employment in language and in logic, there is . . . a sound pragmatic defence. But in metaphysics the concept is sheer error. This error does not consist in the employment of the word 'substance'; but in the employment of the notion of an actual entity which is characterised by essential qualities, and remains numerically one amidst the changes of accidental relations and of accidental qualities. The contrary doctrine is that an ac-

tual entity never changes, and that it is the outcome of whatever can be ascribed to it in the way of quality of relationship. There then remain two alternatives for philosophy: (i) a monistic universe with the illusion of change; and (ii) a pluralistic universe in which 'change' means the diversities among actual entities which belong to some one society of definite type."

6. "The word 'modern' designates two entirely different practices that must remain distinct if they are to remain effective, but have recently begun to be confused. The first set of practices, by 'translation,' creates mixtures between entirely new types of beings, hybrids of nature and culture. The second by 'purification' creates two entirely distinct ontological zones: that of human beings on the one hand; that of non-humans on the other. Without the first set the practices of purification would be pointless. Without the second the work of translation would be slowed down, limited, or even ruled out" (Latour, *We Have Never Been Modern,* 10).

7. See Haraway on the literary understanding of comic as "reconciled, in harmony, secure in the confidence of the restoration of the normal and non-contradictory" (*Modest_Witness@Second_Millennium.FemaleMan©_Meets_Onco Mouse™,* 170).

8. Northrop Frye suggests that comedy is a type of generic plot in literature and drama "At the beginning of the play obstructing characters are usually in charge of the play's society, and the audience recognizes that they are usurpers. At the end of the play the device in the plot that brings hero and heroine together causes a new society to crystallize around the hero . . . The appearance of this new society is signaled by some kind of party. . . . weddings are most common, and sometimes so many of them occur as in the quadruple wedding of *As You Like It,* that they suggest . . . a dance" (*Anatomy of Criticism: Four Essays* [Princeton, NJ: Princeton University Press, 1957], 163).

9. The finding of "the other" in the space of the colonizer is a recognized phenomenon in postcolonial studies.

Chapter Eight

1. I adopted this contrast between complexity of worlds, and the complicatedness that is embedded in objects, from Shirley Strum and Bruno Latour, "Redefining the Social Link: From Baboons to Humans," *Information sur les Sciences Sociales* 26 (1987): 783–802.

2. "Incorporating practices . . . do not exist 'objectively' independently of their being performed [and] are acquired in such a way as to not require explicit reflection on their performance. . . . [T]hey are preserved only through their performance; and because of their performativity and their formalisation they . . . are not easily susceptible to critical scrutiny. . . . [They] contain a measure of insurance against the process of cumulative questioning entailed in all discursive practices. This is the source of their importance and persistence as mnemonic systems. Every group will entrust to bodily automatisms the values and categories which they are most anxious to conserve. They will know how well the past can be kept in mind by a habitual memory sedimented in the body" (Paul Connerton, *How Societies Remember* [Cambridge: Cambridge University Press, 1989], 102).

3. Connerton, *How Societies Remember*, 72.

4. Connerton *(How Societies Remember)* briefly considers this form of incorporated practices in discussing the incorporation of the "seeing" necessary to watch/read a film.

Chapter Nine

1. Watson, "Investigating the Social Foundations of Mathematics."

2. Others have sought to demonstrate that mathematical objects are social constructs. Notable are those working in the tradition of ethnomethodology (Eric Livingston, *The Ethnomethodological Foundations of Mathematics* [London: Routledge and Kegan Paul, 1986]; Eric Livingston, *Making Sense of Ethnomethodology* [London: Routledge and Kegan Paul, 1987], 132–37; Lave, "Values of Quantification") and the rather differently located field of ethnomathematics (Ubiratan D'Ambrosio, *From Reality to Action: Reflections on Education and Mathematics* [São Paulo: Summus Editorial, 1986]; T. N. Carraher, D. W. Carraher, and A. D. Schlieman, "Written and Oral Mathematics," *Journal for Research in Mathematics Education* 18 [1987]: 83–97; Carraher, Carraher, and Schlieman, "Mathematics in the Streets and in the Schools"). Ethnomethodologists seek to expose the ontologies of mathematical objects by detailing the living practices of those using such objects. Those working in ethnomathematics are, for the most part, seeking ways for schools to better facilitate the construction of mathematical objects in learners' minds.

In the reports of ethnomethodologists and ethnomathematicians, we see mathematical objects like natural number being juggled in distinctly different ways in the different discourses in which people use them (Lave, "Values of Quantification"; Nunes Carraher, "Street Mathematics and School Mathematics"). But for all the attention to detail, the mathematical objects themselves remain firmly closed black boxes. The methodology is not up to the task of levering open the boxes, perhaps because the timing is wrong. Unless, of course, ethnomethodologists turn their attention to the negotiations that parents and (some) teachers regularly undertake with children as they train them to use these objects, the critical time framework that would allow natural numbers or angles to be unpacked with the ethnomethodological technique is by now rather remote (Livingston, *Ethnomethodological Foundations of Mathematics*, x). Ethnomethodologists cripple their investigation of the construction of mathematical objects by ruling out historical and cultural analysis (David Bloor, *Wittgenstein: A Social Theory of Knowledge* [London: Macmillan, 1983], 137–38).

In contrast, Bloor thoroughly sociologizes the study of the foundations of mathematics with his comparative theory of language games, seeking "a few, significant variables and then [to] describe the machinery linking these variables to the cultural forms and patterns of verbal exchange that they generate" (Bloor, *Wittgenstein*, 148). Crucial to Bloor's argument are Wittgenstein's notion of family resemblence and the question of the degree of definiteness associated with the use of "significant variables." Bloor is attempting to generate "a theory of *a priori* knowledge."

While Bloor's ingenious investigative machinery does provide us with suggestions about how the conditions of production of mathematical objects might be investigated, he gives us nothing to go on in prizing open the objects themselves. If we are to be convincing in asserting that mathematical objects have been constructed by people as they went about their living as social beings, more than the conditions of their production must be demonstrated. We must be able to show what people have used to accomplish the construction of these objects in their interactions with each other and the material world, and how they have used them.

3. This account of predicating-designating is taken from W. V. O. Quine, *Theories and Things* (Cambridge: Harvard University Press, Belknap Press, 1981) and *The Roots of Reference* (La Salle, IL: Open Court, 1974). In adopting it as a *general* account of predicating-designating, however, I take it that the account can be construed as a nonspecifying rendering of the material world as object and action. It does not necessarily imply that objects will be spatiotemporal objects, or "bodies" as Quine seems inclined to assume. This taking of his account of predicating-designating as a general account means that we seem to disagree over the extent of indeterminacy in translation. In fact, I agree with Quine that the categories generated in any particular form of predicating-designating are inscrutable. Whereas Quine focuses on the impossibility of the process of scrutiny, keeping intact the assumption that there are given categories, I would insist that there are no given categories to scrutinize—the world is not always already sectioned in categories. Thus I agree with Quine on the matter of inscrutability, while insisting on comparative effability of the categories rendered in predicating-designating—the categorizing of one language can be talked about both in it and in another language. I go on here to make just such a comparison of the categories generated in English and Yoruba predicating-designating.

4. This goes back to Ferdinand De Saussure in the original articulation of the structuralist view of language (*Course in General Linguistics,* ed. C. Bally and A. Sechehaye with A. Reidlinger, trans. W. Baskin [London: Peter Owen, 1959], 114).

5. Michael Halliday, *Spoken and Written Language* (Geelong, Australia: Deakin University Press, 1985), 10.

6. In commenting on this invented interrogation, Karin Barber (personal communication, 1986) noted that "it would be most unusual for a Yoruba speaker to actually say "Ọkọ́ ni ó jẹ́" in response to the question "Kí nìyìí?" The normal answer would be "Ọkọ́ ni": the notion of intrinsic characteristics, evoked by the verb *jẹ́* is not actually being continually appealed to; it's not used in all that many contexts.

7. Much of my capacity to make these translations developed when I was a student in a Yoruba language class at the University of Ife, conducted by Prof. Karin Barber. In large part, it was the brilliance of her teaching that enabled the arguments I make here.

The most extensive dictionary (also part grammar and part encyclopedia) is R. C. Abraham, *Dictionary of Modern Yoruba.* Abraham built on the earlier *A Dictionary of the Yoruba Language* (Oxford: Oxford University Press, 1950), which in turn, according to the publisher's note, built on a century of work by Yoruba- and English-speaking members of the Church Missionary Society, particularly Samuel Crowther, E. J. Sowande, Mrs. E. Fry, and T. A. J. Ogunbiyi. Grammars came rather later. First was E. C. Rowlands, *Teach Yourself Yoruba* (London: English

Universities Press, 1969). Later came grammars written by native Yoruba speakers trained in the Western discipline of linguistics. They can all be understood, to use Ayo Bamgbose's term, as "structural grammar[s] of the Yoruba language" (*Short Yoruba Grammar,* viii). They are all thoroughly informed by the foundationalist imaginary of structuralist linguistics. Included in this group are Awobuluyi, *Essentials of Yoruba Grammar;* Bamgbose, *Short Yoruba Grammar;* P. O. Ogunbowale, *The Essentials of Yoruba Grammar* (Ibadan: Hodder and Stoughton, 1970). More recent publications focus on conversation. They include Karin Barber, *Yoruba Dun Un So* (Ibadan: New Horn Press, 1984); Bayo Ogundijo, *Yoruba in Tables* (Birmingham, United Kingdom: Center for West African Studies, 1994).

I have used all these in learning and puzzling about translations between English and Yoruba. As well, I tested the patience of many Yoruba-speaking friends and colleagues, who invariably answered my queries with tolerance and interest. I am deeply indebted to them all.

8. Awobuluyi, *Essentials of Yoruba Grammar,* 96.

9. Abraham, *Dictionary of Modern Yoruba.*

10. Translations of Yoruba language utterances into English are crucial to the argument in this paper. However, only the exceptional reader will be in a position to evaluate the translations. Since the rendering of Yoruba language utterances into English is crucial, I make several attempts in English to catch the subtleties I am trying to focus on. The first translation conveys most precisely what is encoded in the Yoruba language utterance; subsequent glosses work toward less clumsy and more acceptable English.

A reader might well question the veracity of the translations and hence be quite unmoved by the argument. Here the readers must rely on the referees of the paper who speak Yoruba: "The analysis of the Yoruba material is both accurate and profound." Perhaps the knowledge that a panel of Yoruba academics, deeply bilingual with respect to English, have seen fit to publish an earlier version of the paper will also carry some weight (Watson, "Applying Numbers to Nature").

11. In fact, it seems that there are two thresholds: children who are proficient in neither language show a retardation of cognitive growth, but children who are competent users of radically different languages show an acceleration of conceptual learning in both languages. I have demonstrated this second threshold in Yoruba children who are bilingual with respect to English (Watson, "Learning to Apply Numbers to Nature").

12. I use "reasonable" here in the sense that people can reason it out in ways that are generally understood and that use objects that by general agreement are in the world.

13. Numerals appear in four different forms in Yoruba language use. There is a primary numeral form from which the three other forms derive. The origins of the primary set of numerals are ancient, and etymologically their origins are lost. However, they all appear to be elisions of introducers and verbs, nominalized verb phrases. When used in talk to convey multiplicity, the primary nominalized verb phrase form is further modified by elision with the verb *mú* (Awobuluyi, *Essentials of Yoruba Grammar*), an obsolete form of the verb *mún* (to take or pick up several

things in a group or as one). Thus *èji* (in the mode of two) becomes *méjì* (in the mode collected together in the mode two).

14. I am indebted to Mr. A. Laosebikan, my former colleague in the Institute of Education, Obafemi Awolowo University, Ile-Ife, for this account of the units of valuing in Yoruba language.

15. The names of the units of *ẹyọ (ìdì)* were given to me by Mr. A. Adeokun, Institute of African Studies, University of Ibadan.

16. Ifrah, *From One to Zero,* 55.

17. It seems to me that attempts to identify the referents of number words are as misguided as attempts to identify the referent of a part of a ruler. What is important about number names is that they form a sequence. Within any numbering system it is the place marked and how this is achieved that matters.

Chapter Ten

1. Horton, "African Traditional Thought and Western Science," referred to, for example, in Latour, *We Have Never Been Modern,* 42; Kwame Anthony Appiah, *In My Father's House: Africa in the Philosophy of Culture* (London: Methuen, 1992).

2. See Haraway, *Modest_Witness@Second_Millennium.FemaleMan©_Meets_Onco Mouse™,* 295–96 n. 12.

3. Horton, "African Traditional Thought and Western Science," 162.

4. I deal only with the literalization of the first section of the paper. The literalizations in the next two sections are similar to those I have already everted in chapters 4 and 7.

5. Bruno Latour, "Give Me a Laboratory and I Will Raise the World," in *Science Observed,* ed. K. Knorr-Cetina and M. Mulkay (London: Sage, 1983), 141–70.

6. Quine, *Word and Object,* 51.

7. Quine, *Word and Object,* i.

8. Quine, *Word and Object,* 29.

9. Quine, *Word and Object,* 51.

10. It is now obvious that Carnap's positivist characterization of quantification as necessarily proceeding through qualitative "classificatory concepts" assumes that all quantification must begin with spatiotemporal particulars. It is obviously flawed even in its own positivist terms. Only spatiotemporal particulars can be classified on the basis of their qualitative characteristics in the way Carnap defines classificatory concepts. A sortal particular like rabbitmatter cannot be classified in terms of qualitative characteristics, since they are constituted on this basis. They might be placed in classes defined on the basis of modalities; however, modal classification differs from qualitative classification in being simply binary. A sortal particular manifests in a specific mode, or it does not. Qualitative classification admits of degrees or extents of a quality. Quine helps us see an assumption that lies at the base of Carnap's quite orthodox exposition of quantification. It is assumed that the world determinately presents as spatiotemporal particulars—a quite unwarranted assumption, as Quine has so elegantly demonstrated. As far as Quine is concerned, however, we have no way of testing whether this is universally true for any language

community other than our own, so we may as well go along with it. Quine acknowledges the assumption.

11. The outrageous contrivance of such an unlikely harmonization is comic. In another setting, such an impossible reconciliation would provoke laughter in anyone. This analysis of the workings of my text has revealed it as comic in just the way *Twelfth Night* is comic in its contrivances of accomplished resolutions. And the appropriate response in each case is laughter. We see the absurdities, yet there is something in the contrivances of plot, marrying this character off to that, making this and that simultaneously both utterly the same and utterly different, which clicks with us. It is an absurd exaggeration of our intuition that sameness is always difference—the dance of mimesis and alterity.

12. This contrasting pair maps onto "theories of archetypal meaning" in literary studies (see, for example, Frye, *Anatomy of Criticism,* 141) helping us identify how the explanatory power of foundationism is the flip side of the explanatory power of transcendence. In such literary analysis, in tragedy (which I figure as Quine's position) the hero is defeated by forces outside and larger than his world; he must accept the despicable. In comedy (which I figure as my position), a return to harmony and normality is achieved through contrivance, and a new society inaugurated.

Chapter Eleven

1. Understanding acts as generative of tensions, I see myself as drawing on the formulations of Kathryn Pyne Addelson *(Moral Passages).* Pyne Addelson gives a suggestive re-presentation of Dewey's insights, drawing us back to a classic 1896 paper by Dewey, which critiques psychologists' construction of the reflex arc (John Dewey, "The Unit of Behaviour [The Reflex Arc Concept in Psychology]," *Psychological Review* 3, no. 4 [1896]: 357–70).

In that classic paper, Dewey refused to accept the dualism of stimulus and response that was becoming influential among American psychologists and instead took the unit of experience to be the act. It is out of the act that stimulus and response, subject and object, are generated.

In the orthodox, foundationist interpretation of the psychology experiment that is the subject of Dewey's critique, a candle flame is an object in the world with the power to stimulate. It causes a child to reach. Similarly, the pain the child experiences is an object in the inner world with power to cause the child to pull back, the reflex action. There, the meaning of the candle flame and the pain is taken to be given, independent of the child. The "outside existence of these entities" is taken to be the origin of all certainty in meaning making and particularly in language use.

In contrast, Dewey points to complex, mutually engendered "condensations" of baby as self or subject and as object, and of flame too; both subjects and objects are outcomes, participants in collective acting. It is in the circuit of the act that various meanings stabilize. Pyne Addelson significantly extends Dewey's interpretations pointing to other outcomes of this collective acting, involving the flame as much as the child, and the parents, the experimental pyschologist and the laboratory, and so on. In particular, she points to time as an outcome, along with all those other outcomes that Dewey identified: "the act has a lived time, in which the child first reached for a bright object, and then learned it was a painful one. The reaching is the past for the drawing back. All of this is within the circuit of the act itself" (144–46).

2. Brian O'Shaughnessy, "Proprioception and the Body Image," in *The Body and the Self*, ed. Jose Luis Bermudez, Anthony Marcel, and Naomi Eilan (Cambridge: MIT Press, 1995), 175–203.

3. O'Shaughnessy, "Proprioception and the Body Image," 176.

4. O'Shaughnessy, "Proprioception and the Body Image," 190.

5. This point is the start of Wittgenstein's exploration of the logic of first-person psychological sentences. His showing that the concept of verification has no application to some first-person psychological sentences is my starting point (Ludwig Wittgenstein, *Philosophical Investigations* [Oxford: Basil Blackwell, 1958]). Those sections following paragraph 243 of the *Investigations* are usually thought of as "the private language argument." Following Saul Knipke (*Wittgenstein on Rules and Private Language* [Oxford: Basil Blackwell, 1982], viii), I think the importance of those paragraphs should not be isolated from those that precede it.

6. The notion that some sentences are "expressions" rather than "propositions," "statements," or "thoughts" was first applied by Wittgenstein to psychological sentences in the first person, utterances like "I have a toothache." Adults reach a child to replace her moans and the holding of the side of her face with words and sentences like "Toothache!" They are teaching her new pain behavior (Wittgenstein, *Philosophical Investigations*, para. 244). The articulation is no more due to thinking or reasoning, and I would add no more the act of a little girl alone, than the preverbal behaviors of crying and biting on a teething rusk. Wittgenstein calls these verbal utterances *Ausserungen* to indicate that they are immediate expressions of familiarity, pain, discomfort, surprise, outrage, and so on, and are not the result of thought; they are not propositions or statements. I adopt Malcolm's translation of *Ausserungen* as "expression" (Norman Malcolm, *Wittgenstein: Nothing Is Hidden* [Oxford: Basil Blackwell, 1986], chapter 8).

7. "Surely that this way of behaving is *prelinguistic*: that a language game is based on it, that it is the prototype of a way of thinking and not the result of thinking" (Ludwig Wittgenstein, *Zettel*, ed. G. E. M. Anscombe and G. H. von Wright, trans. G. E. M. Anscombe [Oxford: Basil Blackwell, 1967], para. 541).

8. Ludwig Wittgenstein, "Cause and Effect: Intuitive Awareness," manuscript notes, ed. Rush Rhees, trans. Peter Winch, *Philosophia* 6, nos. 3–4 (1976), 416.

9. See Jane Guyer, "Diversity at Different Levels: Farms and Community in Western Nigeria," *Africa* 66 (1996): 71–89. This article is included in an edition of the journal devoted to exploring the "social shaping" of the concept of biodiversity in Africa. Jane Guyer and Paul Richards, the editors, suggest that there is a need to understand how the concept of biodiversity might be influenced by local understandings of diversity, in its employment in African contexts.

10. This is the puzzle of "levels" in Guyer's paper "Diversity at Different Levels."

References

Abraham, R. C. *Dictionary of Modern Yoruba*. London: Hodder and Stoughton, 1962.

Afolabi Ojo, G. J. *Yoruba Culture: A Geographical Analysis*. London: University of London Press, 1966.

Althusser, Louis. *Lenin and Philosophy and Other Essays*. Trans. Ben Brewster. New York: Monthly Review Press, 1971.

Appadurai, Arjun. "Number in the Colonial Imagination." In *Orientalism and the Postcolonial Predicament: Perspectives on South Asia,* ed. Carol Breckenridge and Peter van der Veer. Philadelphia: University of Pennsylvania Press, 1993.

Appiah, Kwame Anthony. *In My Father's House: Africa in the Philosophy of Culture*. London: Methuen, 1992.

Armstrong, R. G. *Yoruba Numerals*. Ibadan: Oxford University Press, 1962.

Audi, R., ed. *The Cambridge Dictionary of Philosophy*. Cambridge: Cambridge University Press, 1995.

Awe, Bolanle. "The Ajele System: A Study of Ibadan Imperialism in the Nineteenth Century." *Journal of the Historical Society of Nigeria* 3 (1964): 47–71.

Awobuluyi, P. Oladele. *Essentials of Yoruba Grammar*. Ibadan: Oxford University Press, 1978.

Bamgbose, Ayo. *A Short Yoruba Grammar*. 2nd edition. Ibadan: Heinemann, 1974.

Barber, C. L. *The Story of English*. London: Pan Books, 1972.

Barber, Karin. *Yoruba Dun Un So*. Ibadan: New Horn Press, 1984.

———. *I Could Speak until Tomorrow: Oriki, Women, and the Past in a Yoruba Town*. Edinburgh: Edinburgh University Press, 1991.

Barnes, B. *Interests and the Growth of Knowledge*. London: Routledge and Kegan Paul, 1977.

Benjamin, Walter. "Doctrine of the Similar." *New German Critique* 17 (spring 1979): 65–69.

Berry, J. W., and P. R. Dasen, eds. *Culture and Cognition: Readings in Cross-Cultural Psychology.* London: Methuen, 1974.

Bloor, David. *Knowledge and Social Imagery.* 2nd edition. Chicago: University of Chicago Press, 1991.

———. *Wittgenstein: A Social Theory of Knowledge.* London: Macmillan, 1983.

Bowen, T. J. *Grammar and Dictionary of the Yoruba Language.* Washington, DC: Smithsonian Institution, 1858.

Bruner, J. S., R. R. Oliver, and P. M. Greenfield, eds. *Studies in Cognitive Growth.* New York: Wiley, 1966.

Burrow, J. W. *The Crisis of Reason: European Thought, 1848–1914.* New Haven: Yale University Press, 2000.

Butler, Judith. *Excitable Speech: A Politics of the Performative.* New York: Routledge, 1997.

———. *The Psychic Life of Power: Theories in Subjection.* Stanford: Stanford University Press, 1997.

Callon, Michel. "Some Elements of a Sociology of Translation: Domestication of the Scallops and the Fishermen of St. Brieuc Bay." In *Power, Action, and Belief: A New Sociology for Knowledge?* ed. John Law, Sociological Review Monograph 32. London: Routledge and Kegan Paul, 1986.

———. "Techno-economic networks and irreversibility." In *A Sociology of Monsters: Essays on Power, Technology, and Domination,* ed. John Law, Sociological Review Monograph 38. London: Routledge, 1991.

———. "Actor-Network Theory: The Market Test." In *Actor Network Theory and After,* ed. John Law and John Hassard. Oxford: Blackwell, 1999.

Callon, Michel, John Law, and Arie Rip, eds. *Mapping the Dynamics of Science and Technology: Sociology of Science in the Real World.* London: Macmillan, 1986.

Carnap, Rudolf. *Philosophical Foundations of Physics: An Introduction to the Philosophy of Science.* Ed. Martin Gardner. Chicago: University of Chicago Press, 1966.

Carraher, T. N., D. W. Carraher, and A. D. Schliemann. "Mathematics in the Streets and in the Schools." *British Journal of Developmental Psychology* 3 (1985): 21–29.

———. "Written and Oral Mathematics." *Journal for Research in Mathematics Education* 18 (1987): 83–97.

Carter, Michael. "Notes on Imagination, Fantasy, and the Imaginary." In *Imaginary Materials: A Seminar with Michael Carter,* ed. John Macarthur. Brisbane: Institute of Modern Art, 2000.

Castoriadis, Cornelius. *The Imaginary Institution of Society.* Cambridge: MIT Press, 1987.

Code, Lorraine. *Epistemic Responsibility.* Hanover, NH: University Press of New England, 1987.

———. *Rhetorical Spaces: Essays on Gendered Locations.* New York: Routledge, 1995.

Cole, M. *Cultural Psychology: A Once and Future Discipline.* Cambridge: Harvard University Press, Belknap Press, 1996.

Cole, M., and S. Scribner. *Culture and Thought: A Psychological Introduction.* New York: Wiley, 1974.

Conant, Levi Leonard. *The Number Concept: Its Origins and Development.* New York: Macmillan and Co., 1896.

Connerton, Paul. *How Societies Remember.* Cambridge: Cambridge University Press, 1989.

Crapanzano, Vincent. *Serving the Word: Literalism in America from the Pulpit to the Bench.* New York: New Press, 2000.

Crowther, Samuel Adjai. *Vocabulary of the Yoruba Language.* London: Church Missionary Society, 1843.

Crump, T. *The Anthropology of Numbers.* Cambridge: Cambridge University Press, 1990.

D'Ambrosio, Ubiratan. *From Reality to Action: Reflections on Education and Mathematics.* São Paulo: Summus Editorial, 1986.

Dantzig, Tobias. *Number: The Language of Science.* New York: Free Press, 1954.

Davis, P. J., and R. Hersh. *The Mathematical Experience.* London: Pelican Books, 1983.

De Saussure, Ferdinand. *Course in General Linguistics.* Ed. C. Bally and A. Sechehaye with A. Reidlinger. Trans. W. Baskin. London: Peter Owen, 1959.

Dewey, John. "The Unit of Behaviour (The Reflex Arc Concept in Psychology)." *Psychological Review* 3, no. 4 (1896): 357–70.

Ekundayo, S. A. "Aspects of Underlying Representations in the Yoruba Noun Phrase." Ph.D. thesis, University of Edinburgh, 1972.

———. "Vigesimal Numeral Derivational Morphology: Yoruba Grammatical Competence Epitomized." Paper presented in the Linguistics Department, University of Ife, 1975. Mimeo.

Fadipe, N. A. *The Sociology of the Yoruba.* 1939. Reprint, Ibadan: University of Ibadan Press, 1970.

Falola, Toyin. "In Praise of *Oriki.*" *Journal of African History* 33, no. 2 (1992): 331–32.

Finnegan, Ruth, and Robin Horton, eds. *Modes of Thought.* London: Faber, 1973.

Frege, Gottlob. *The Foundations of Arithmetic.* Trans. J. L. Austin. Oxford: Basil Blackwell, 1974.

Frye, Northrop. *Anatomy of Criticism: Four Essays.* Princeton, NJ: Princeton University Press, 1957.

Gay, J., and M. Cole. *The New Mathematics and an Old Culture: A Study of Learning among the Kpelle of Liberia.* New York: Holt, Rinehart and Winston, 1967.

Geertz, Clifford. *The Interpretation of Cultures.* New York: Basic Books, 1973.

Gellner, Ernest. *Legitimation of Belief.* Cambridge: Cambridge University Press, 1974.

Giere, Ronald, and Alan Richardson, eds. *Minnesota Studies in the Philosophy of Sci-*

ence. Vol. 16, *Origins of Logical Empiricism*. Minneapolis: University of Minnesota Press, 1996.

Gödel, Kurt. *On Formally Undecidable Propositions*. New York: Basic Books, 1962.

Greenfield, P. M. "On Culture and Conservation." In *Studies in Cognitive Growth*, ed. J. S. Bruner, R. R. Oliver, and P. M. Greenfield. New York: Wiley, 1966.

Guyer, Jane. "Diversity at Different Levels: Farms and Community in Western Nigeria." *Africa* 66 (1996): 71–89.

Hacking, Ian. "Language, Truth, and Reason." In *Rationality and Relativism*, ed. Martin Hollis and Steven Lukes. Cambridge: MIT Press, 1982.

———. *Representing and Intervening: Introductory Topics in the Philosophy of Science*. Cambridge: Cambridge University Press, 1983.

Hall, Stuart. "When Was the 'Post-Colonial'? Thinking at the Limit." In *The Post-Colonial Question: Common Skies, Divided Horizons*, ed. Iain Chambers and Lidia Curti. London: Routledge, 1996.

Hallen, Barry. "Robin Horton on Critical Philosophy and Traditional Religion." *Second Order* 6, no. 1 (1977): 81–92.

Hallen, Barry, and J. O. Sodipo. *Knowledge, Belief, and Witchcraft*. London: Ethnographica, 1986.

Halliday, Michael. *Spoken and Written Language*. Geelong, Australia: Deakin University Press, 1985.

Hallpike, C. R. *The Foundations of Primitive Thought*. Oxford: Clarendon Press, 1979.

Haraway, Donna. *Modest_Witness@Second_Millennium.FemaleMan©_Meets_OncoMouse™*. New York: Routledge, 1997.

Hazewinkel, M., ed. *Encyclopaedia of Mathematics*. Dordrecht: Kluwer Academic, 1993.

Hofstadter, Douglas. *Gödel, Escher, Bach: An Eternal Golden Braid*. Harmondsworth: Penguin Books, 1980.

Hollis, Martin, and Steven Lukes, eds. *Rationality and Relativism*. Cambridge: MIT Press, 1982.

Honderich, T., ed. *Oxford Companion to Philosophy*. Oxford: Oxford University Press, 1995.

Horton, Robin. "African Traditional Thought and Western Science." *Africa* 37 (1967): 50–71, 155–87.

Ifrah, Georges. *From One to Zero: A Universal History of Numbers*. New York: Viking Penguin, 1985.

Johnson, Samuel. *The History of the Yorubas from the Earliest Times to the Beginning of the British Protectorate*. Lagos: CMS (Nigeria) Bookshops, 1921.

Kripke, Saul A. *Wittgenstein on Rules and Private Language*. Oxford: Basil Blackwell, 1982.

Kuhn, Thomas. *The Structure of Scientific Revolutions*. Chicago: University of Chicago Press, 1962.

Latour, Bruno. "Give Me a Laboratory and I Will Raise the World." In *Science Observed*, ed. K. Knorr-Cetina and M. Mulkay. London: Sage, 1983.

———. *Science in Action: How to Follow Scientists and Engineers through Society.* Milton Keynes, United Kingdom: Open University Press, 1987.

———. *The Pasteurization of France.* Part 1. Cambridge: Harvard University Press, 1988.

———. *We Have Never Been Modern.* Trans. Catherine Porter. Cambridge: Harvard University Press, 1993.

———. "The 'Pédofil' of Boa Vista." *Common Knowledge* 4 (1995): 144–87.

———. "Do Scientific Objects Have a History? Pasteur and Whitehead in a Bath of Lactic Acid." *Common Knowledge* 5 (1996): 70–89.

———. "A Few Steps toward an Anthropology of the Iconoclastic Gesture." *Science in Context* 10 (1997): 63–83.

———. "From the World of Science to the World of Research." *Science* 280 (April 1998): 207–8.

———. "On Recalling ANT." In *Actor Network Theory and After,* ed. John Law and John Hassard. Oxford: Blackwell, 1999.

———. *Pandora's Hope: Essays on the Reality of Science Studies.* Cambridge: Harvard University Press, 1999.

Latour, Bruno, and Steve Woolgar. *Laboratory Life: The Construction of Scientific Facts.* Beverly Hills: Sage Publications, 1979.

Lave, Jean. "The Values of Quantification." In *Power, Action, and Belief: A New Sociology of Knowledge,* ed. John Law, Sociological Review Monograph 32. London: Routledge and Kegan Paul, 1986.

Law, John. "On the Methods of Long Distance Control: Vessels, Navigation, and the Portuguese Route to India." In *Power, Action, and Belief: A New Sociology for Knowledge?* ed. John Law, Sociological Review Monograph 32. London: Routledge and Kegan Paul, 1986.

———. Introduction to *A Sociology of Monsters: Essays on Power, Technology, and Domination,* ed. John Law, Sociological Review Monograph 38. London: Routledge, 1991.

———. *Organising Modernity.* Oxford: Blackwell, 1994.

———. "After ANT: Complexity, Naming, and Topology." In *Actor Network Theory and After,* ed. John Law and John Hassard. Oxford: Blackwell, 1999.

Law, John, and Michel Callon. "The Life and Death of an Aircraft: A Network Analysis of Technical Change." In *Shaping Technology—Building Society: Studies in Sociotechnical Change,* ed. Weibe Bijker and John Law. Cambridge: MIT Press, 1992.

Livingston, Eric. *The Ethnomethodological Foundations of Mathematics.* London: Routledge and Kegan Paul, 1986.

———. *Making Sense of Ethnomethodology.* London: Routledge and Kegan Paul, 1987.

Lloyd, Barbara. "Studies in Conservation with Yoruba Children of Differing Ages and Experience." *Child Development* 42 (1971): 415–28.

———. "Cognitive Development, Education, and Social Mobility." In *Universals of Human Thought: Some African Evidence,* ed. Barbara Lloyd and John Gay. Cambridge: Cambridge University Press, 1981.

Lloyd, Barbara, and Brian Easton. "The Intellectual Development of Yoruba Children." *Journal of Cross Cultural Psychology* 8 (1977): 3–15.

Lovejoy, A. O. *The Great Chain of Being.* Cambridge: Harvard University Press, 1936.

———. *Essays in the History of Ideas.* Baltimore: Johns Hopkins Press, 1948.

Malcolm, Norman. *Wittgenstein: Nothing Is Hidden.* Oxford: Basil Blackwell, 1986.

Mann, Adolphus. "Notes on the Numeral System of the Yoruba Nation." *Journal of the Royal Anthropological Institute* 16 (1887): 59–64.

Menninger, Karl. *Number Words and Number Symbols: A Cultural History of Numbers.* Cambridge: MIT Press, 1969.

Mimica, Jadran. *Intimations of Infinity: The Cultural Meanings of the Iqwaye Counting and Number System.* Oxford: Berg, 1992.

Mol, Annemarie. "Missing Links, Making Links: The Performance of Some Atheroscleroses." In *Differences in Medicine: Unravelling Practices, Techniques, and Bodies,* ed. Marc Berg and Annemarie Mol. Durham: Duke University Press, 1998.

———. "Ontological Politics: A Word and Some Questions." In *Actor Network Theory and After,* ed. John Law and John Hassard. Oxford: Blackwell, 1999.

Mol, Annemarie, and Marc Berg. "Principles and Practice of Medicine: The Co-existence of Various Anaemias." *Culture, Medicine, and Psychiatry* 18 (1994): 247–65.

Mol, Annemarie, and John Law. "Situated Bodies and Distributed Selves: On Doing Hypoglycaemia." Paper presented to Netherlands Graduate School of Science, Technology, and Modern Culture/Centre de Sociologie de l'Innovation Conference on the Body, Paris, September 1998.

Musgrove, Frank. *Education and Anthropology: Other Cultures and the Teacher.* Chichester: John Wiley and Sons, 1982.

Ogunbowale, P. O. *The Essentials of Yoruba Grammar.* Ibadan: Hodder and Stoughton, 1970.

Ogundijo, Bayo. *Yoruba in Tables.* Birmingham, United Kingdom: Centre for West African Studies, 1994.

O'Shaughnessy, Brian. "Proprioception and the Body Image." In *The Body and the Self,* ed. Jose Luis Bermudez, Anthony Marcel, and Naomi Eilan. Cambridge: MIT Press, 1995.

Peel, J. D. Y. *Ijeshas and Nigerians: The Incorporation of a Yoruba Kingdom, 1890s–1970s.* Cambridge: Cambridge University Press, 1983.

———. "The Cultural Work of Yoruba Ethnogenesis." In *History and Ethnicity,* ed. Elizabeth Tonkin, Maryon McDonald, and Malcolm Chapman. London: Routledge, 1989.

Popper, Karl. *Conjectures and Refutations: The Growth of Scientific Knowledge.* New York: Basic Books, 1962.

Porter, Theodore. *Trust in Numbers: The Pursuit of Objectivity in Science and Public Life.* Princeton: Princeton University Press, 1995.

Price-Williams, Rhys. "A Study concerning Concepts of Conservation of Quantities among Primitive Children." *Acta Psychologica* 18 (1961): 297.

Pyne Addelson, Kathryn. "Knowers/Doers and Their Moral Problems." In *Feminist Epistemologies,* ed. Linda Alcoff and Elizabeth Potter. New York: Routledge, 1993.

———. *Moral Passages: Towards a Collectivist Moral Theory.* New York: Routledge, 1994.

Quine, W. V. O. *Word and Object.* Cambridge: MIT Press, 1960.

———. *The Roots of Reference.* La Salle, IL: Open Court, 1974.

———. *Theories and Things.* Cambridge: Harvard University Press, Belknap Press, 1981.

Rains, Prudence. *Becoming an Unwed Mother: A Sociological Account.* Chicago: Aldine Atherton, 1971.

Raymond, Eric S. *The New Hacker's Dictionary.* Cambridge: MIT Press, 1993.

Reed, H. J., and Jean Lave. "Arithmetic as a Tool for Investigating Relations between Culture and Cognition." *American Ethnologist* 6 (1979): 568–82.

Rotman, Brian. *Signifying Nothing: The Semiotics of Zero.* London: Macmillan, 1987.

Rouse, Joseph. *Knowledge and Power: Towards a Political Philosophy of Science.* Ithaca, NY: Cornell University Press, 1987.

Rowlands, E. C. *Teach Yourself Yoruba.* London: English Universities Press, 1969.

Said, E. W. *Orientalism.* New York: Vintage Books, 1978.

Serres, Michel. *Hermes: Literature, Science, Philosophy.* Ed. Josue V. Harari and David F. Bell. Baltimore: Johns Hopkins University Press, 1982.

———. *Detachment.* Trans. Genevieve James and Raymond Federman. Athens: Ohio University Press, 1989.

———. *Angels: A Modern Myth.* Ed. Philippa Hurd. Paris: Flammarion, 1995.

———. *Genesis.* Trans. Genevieve James and James Nielson. Ann Arbor: University of Michigan Press, 1995.

Soyinka, Wole. *The Open Sore of a Continent: A Personal Narrative of the Nigerian Crisis.* Oxford: Oxford University Press, 1996.

Stepan, Nancy. *The Idea of Race in Science: Great Britain, 1800–1960.* London: Macmillan, 1982.

Strathern, Marilyn. *The Gender of the Gift: Problems with Women and Problems with Society in Melanesia.* Berkeley: University of California Press, 1988.

———. *After Nature: English Kinship in the Late Twentieth Century.* Cambridge: Cambridge University Press, 1992.

———. "The Decomposition of an Event." *Cultural Anthropology* 7, no. 7 (1992): 244–54.

———. *Reproducing the Future: Anthropology, Kinship, and the New Reproductive Technologies.* New York: Routledge, 1992.

———. *The Relation Issues in Complexity and Scale.* Cambridge: Prickly Pear Press, 1995.

———. "Environments Within: An Ethnographic Commentary on Scale." Linacre lectures 1996–97, Linacre College, Oxford.

———, ed. *Shifting Contexts: Transformations in Anthropological Knowledge.* New York: Routledge, 1995.

Strum, Shirley, and Bruno Latour. "Redefining the Social Link: From Baboons to Humans." *Information sur les Sciences Sociales* 26 (1987): 783–802.

Taussig, Michael. *Mimesis and Alterity: A Particular History of the Senses.* New York: Routledge, 1993.

Thomas, Nicholas. *Entangled Objects: Exchange, Material Culture, and Colonialism in the Pacific.* Cambridge: Harvard University Press, 1991.

Toy, C. H. "The Yoruban Language." *Transactions of the American Philological Association* 9 (1878): 19–38.

Turkle, Sherry. *The Second Self: Computers and the Human Spirit.* New York: Simon and Schuster, 1984.

Turnbull, David. *Maps Are Territories: Science Is an Atlas: A Portfolio of Exhibits.* Chicago: University of Chicago Press, 1993.

Urton, Gary. *The Social Life of Numbers: A Quechua Ontology of Numbers and Philosophy of Arithmetic.* Austin: University of Texas Press, 1997.

Verran, Helen. "Staying True to the Laughter of Nigerian Classrooms." In *Actor Network Theory and After,* ed. John Law and John Hassard. Oxford: Blackwell, 1999.

———. "Accounting Mathematics in West Africa: Some Stories of Yoruba Number." In *Mathematics across Cultures: The History of Non-Western Mathematics,* ed. Helaine Selin and Ubiritan D'Ambrosio. Dordrecht: Kluwer Academic, 2000.

———. "Logics and Mathematics: Challenges Arising in Working across Cultures." In *Mathematics across Cultures: The History of Non-Western Mathematics,* ed. Helaine Selin and Ubiritan D'Ambrosio. Dordrecht: Kluwer Academic, 2000.

Watson, Helen. "Applying Numbers to Nature: A Comparative View in English and Yoruba." *Journal of Cultures and Ideas* (Ile-Ife, Nigeria) 2–3 (1986): 1–19.

———. "Learning to Apply Numbers to Nature: A Comparison of English Speaking and Yoruba Speaking Children Learning to Quantify." *Educational Studies in Mathematics* 18 (1987): 339–57.

———. "A Wittgensteinian View of Mathematics: Implications for Teachers." In *School Mathematics,* ed. N. Ellerton and K. Clements. Geelong, Australia: Deakin University Press, 1989.

———. "Investigating the Social Foundations of Mathematics: Natural Number in Culturally Diverse Forms of Life." *Social Studies of Science* 20 (1990): 283–312.

Watson, Ruth. *"Civil Disorder Is the Disease of Ibadan": Chieftaincy and Civic Culture in a Colonial City.* London: James Currey, 2001.

Watson Verran, Helen, and David Turnbull. "Science and Other Indigenous Knowledge Systems." In *Handbook of Science and Technology Studies,* ed. Sheila Jasanoff, Gerald Markle, James Petersen, and Trevor Pinch. Thousand Oaks, CA: Sage, 1995.

Watts, Michael. "Nature as Artifice and Artifact." In *Remaking Reality: Nature at the Millennium,* ed. Bruce Braun and Noel Castree. London: Routledge, 1997.

Whitehead, A. N. *Process and Reality: An Essay in Cosmology.* New York: Free Press, 1957.

Wilson, Bryan, ed. *Rationality.* Oxford: Basil Blackwell, 1970.

Winch, P. "Understanding a Primitive Society." *American Philosophical Quarterly* 1 (1964): 307–24.

Wiredu, Kwasi. *Philosophy and an African Culture.* Cambridge: Cambridge University Press, 1980.

Wittgenstein, Ludwig. 1958. *Philosophical Investigations.* Oxford: Basil Blackwell, 1958.

———. *Zettel.* Ed. G. E. M. Anscombe and G. H. von Wright. Trans. G. E. M. Anscombe. Oxford: Basil Blackwell, 1967.

———. "Cause and Effect: Intuitive Awareness." Manuscript notes by Wittgenstein. Ed. Rush Rhees. Trans. Peter Winch. *Philosophia* 6, nos. 3–4 (1976): 410–21.

Zavlavsky, C. *Africa Counts: Number and Pattern in African Culture.* Boston: Prindle, Weber, and Schmidt, 1973.

Index

In references to endnotes, if the subject discussed in the endnote is not directly identified in the text, the page on which the note reference occurs is given in parentheses.

CPSIA information can be obtained
at www.ICGtesting.com
Printed in the USA
BVHW04s2127120818
524137BV00029B/316/P